Fieldwork Ready:
An Introductory Guide to Field Research for
Agriculture, Environment, and Soil Scientists

Fieldwork Ready

An Introductory Guide to Field Research for Agriculture, Environment, and Soil Scientists

Sara E. Vero

Editorial Correspondence:
American Society of Agronomy, Inc.
Crop Science Society of America, Inc.
Soil Science Society of America, Inc.
5585 Guilford Road, Madison, WI 53711-58011, USA

agronomy.org
crops.org
soils.org

Registered Office
John Wiley & Sons, Inc., 111 River Street, Hoboken, NJ 07030, USA

For details of our global editorial offices, customer services, and more information about Wiley products, visit us at www.wiley.com.

Wiley also publishes its books in a variety of electronic formats and by print-on-demand. Some content that appears in standard print versions of this book may not be available in other formats.

Library of Congress Cataloging-in-Publication Data

Name: Vero, Sara, author.
Title: Fieldwork ready : an introductory guide to field research for
   agriculture, environment, and soil scientists / Sara E. Vero.
Description: Hoboken, NJ, USA : Wiley-ACSESS, 2021. | Includes
   bibliographical references and index.
Identifiers: LCCN 2020028489 | ISBN 9780891183754 (paperback)
Subjects: LCSH: Environmental sciences–Research–Methodology. |
   Agriculture–Research–Methodology. | Fieldwork (Educational method)
Classification: LCC GE70 .V47 2020 | DDC 550.72/3–dc23
LC record available at https://lccn.loc.gov/2020028489

doi:10.2134/fieldwork

Cover Design: Wiley
Cover Image: © Jaclyn Fiola

Set in 9.5/12.5pt STIXTwoText by SPi Global, Pondicherry, India

10 9 8 7 6 5 4 3 2 1

With gratitude to all those friends who have dug through the earth, marched across grasslands or waded upstream with me. What a wonderful adventure.

To the reader; my grandfather Paddy Vero said that "An ounce of help is worth a ton of pity." I have been the happy beneficiary of many kind helpers. I hope this book will give you at least that ounce of help when you need it.

"Whatsoever your hand finds to do, do it with all your might"

Ecclesiastes 9.10

Contents

List of Photo Contributors

Jaclyn Fiola
Robert Collins
Jesse Nippert
Nicolette Roach
Krista Keels
Karen Vaughan
Katie O'Reilly
Julie Campbell
Rachel Murphy
Derek Gibson

Brandon Forsythe
Lauren McPhillips
Sian Green
Andrea Brookfield
Sophie Sherriff
Noemi Nazsarkowski
David Jaramillo
Giulia Bondi
Teagasc (Raymond Kelly – Head of Research
    Support)

Preface

Field-based research is a cornerstone of agronomic and environmental science, yielding information that helps us produce crops efficiently, manage resources, and steward the environment. For students and researchers, it allows insights into the real world, which cannot be achieved in the classroom or library alone. Fieldwork is, for many, an exciting and engaging part of their work and studies. However, it brings unique challenges pertaining to experimental design, planning, safety, and team management, in addition to the scientific techniques being employed. The field researcher needs to be well-rounded and adaptable; able to deal with the unexpected and to improvise in response to challenges arising outside of the clean, controlled environment of the laboratory.

Fieldwork Ready is intended to help you to become an effective researcher, whether you are involved in agronomy, soil science, hydrology, geography, or any other field-based study. This book includes advice on design, planning, and logistics, which are essential for all field researchers, and then discusses basic techniques related to environmental monitoring, and soil, water, plant, and wildlife research that any investigator should be familiar with. These are intended as a guide upon which you can and should build further skills. For those of you who are already experienced in the field, this book should help you think more deeply about how and why you do field research, and hopefully, to improve upon your skills and knowledge.

Acknowledgments

Thanks to Danielle Lynch for her patient editing and Karen Brey for creating such wonderful illustrations. I am grateful to each of the reviewers for their intelligent and constructive advice, and for sharing their experience of the field. Special thanks to everyone who so generously donated their fieldwork photos; Vikram Baliga, Giulia Bondi, Andrea Brookfield, Julie Campbell, Bo Collins, Jaclyn Fiola, Brandon Forsythe, Derek Gibson, David Jaramillo, Krista Keels, Colum Kennedy, Juliette Marie, Lauren McPhillips, Rachel Murphy, Jesse Nippert, Katherine O'Reilly, Nikki Roach, Sophie Sherriff, and Karen Vaughan. It has been a pleasure to see your exciting work.

1

Introduction

This manual provides simple guidance to help you perform safe and successful fieldwork as part of your research. The "field" can be urban, rural, or wild. You may work alone or in a team. The experiment may be structured or survey-based in design. You may operate adjacent to your research center or in remote locations. Regardless, there are principles and considerations that can be universally applied that will allow you to implement a robust and meaningful research project and collect quality data. While this manual can help anyone involved in outdoor research, it is particularly aimed toward graduate and undergraduate students, and early-career researchers who are honing their skills and gaining experience. Everyone makes mistakes in their early development, and fieldwork often involves a steep learning curve, potentially hazardous or challenging conditions, and considerable time and financial commitments. Naturally, your unique field of study will determine some of the technical skills that you will build and depend on, but elements of planning, site description, logistics, and teamwork are universal. Experience is the best teacher, but hopefully this manual will help you to make a good start.

What is "Fieldwork?"

Fieldwork is any research or data gathering conducted outdoors, outside of the laboratory, library, or office settings. As researchers, our individual fields can be almost anywhere (Figs. 1.1 and 1.2). For a sociologist, it might be a school, a shopping mall, or wherever there are people. For a marine biologist, it may be on or even deep within the ocean. This particular guide is generally intended for students and researchers in the broad disciplines of soil, crop, and environmental sciences. However, many of the principles discussed throughout this book will be helpful for any researcher venturing outside of the laboratory setting.

For simplicity, I will refer to all outdoor research as "fieldwork" and all indoor research (be it laboratory, desk, or workshop) as "labwork."

The challenge faced by researchers in the field is to apply scientific methodologies into environments which are by their very nature, heterogeneous and subject to limited human control. As field researchers, we cannot control the weather, the movements of wildlife, heterogeneity of soils, rock, or vegetation, and innumerable other factors which may influence the results of our

Fieldwork Ready: An Introductory Guide to Field Research for Agriculture, Environment, and Soil Scientists,
First Edition. Sara E. Vero.
© 2021 American Society of Agronomy, Inc., Crop Science Society of America, Inc., and Soil Science Society of America, Inc. Published 2021 by John Wiley & Sons, Inc.
doi:10.2134/fieldwork.c1

Fig. 1.1 Researchers investigating a soil pit in Ireland. *Source:* Sara Vero.

Fig. 1.2 Field research can take you to some breathtaking scenery. *Source:* Bo Collins.

investigations. This may seem contrary to the scientific method, which typically controls variables and factors so that one or a few factors of particular interest may be examined independently. In reality, outside of the laboratory, these conditions rarely, if ever, exist (Fig. 1.3). Fieldwork is therefore critical to examine how the theories, devices, and processes developed under controlled conditions perform in reality.

Research can be considered to take place within a "hierarchy of complexity" (Read, 2003). Studies that are reductionist in approach, dealing with only one of the many variables which simultaneously influence biological, physical, and chemical functioning in reality, can provide insight into the underlying mechanisms of behavior. However, these effects might be difficult to discern or become less influential at the field scale. These studies offer a high level of "precision," but perhaps, a lower level of "relevance." Conversely, field studies allow a broader understanding of patterns and effects within a "real-world" context. In other words, they have a lower level of "precision," but a high level of "relevance" (Read, 2003). Of course, there is no strict rule regarding this; rather, it is a spectrum along which various experimental approaches are positioned. For this reason, coupled field and lab studies can be used to develop a more integrated understanding. This is common, especially when developing a thesis at graduate level. Let us take an example.

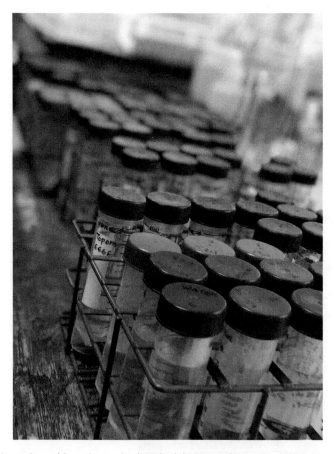

Fig. 1.3 The field is rarely as tidy and organised as the laboratory. *Source:* Bo Collins.

A student investigating potassium (K) requirements of mixed species grassland might conduct three structured experiments.

1) A soil incubation study in the laboratory to indicate the release and adsorption potential. This would indicate fundamental chemical behavior of the soil, without any confounding factors.
2) A pot study in a glasshouse or growth chamber to examine the response to various K levels in different species mixtures (Fig. 1.4). This would give an indicator of the potential implications of K availability.
3) A plot study at field scale over three years (Fig. 1.5). This would reveal the impacts of the behaviors observed in detail in the first two experiments, but at an applied spatial and temporal scale. Results from this approach can be used to develop recommendations for farm management.

While the conclusions from experiment one could be extrapolated to the field scale, without the bridging provided by experiment two, and the real-world implications observed in the field during experiment three, any recommendations derived thus would be vulnerable to overemphasis or misinterpretation. Conversely, while field experiments might reveal the implications or applications of (for example) farm management practices, they may struggle to differentiate the underlying causative factors. A joint approach incorporating both laboratory and field elements can often yield a more comprehensive understanding, and justified conclusions than either can in isolation.

Fig. 1.4 A pot study in a glasshouse can be highly controlled. *Source:* Bo Collins.

Fig. 1.5 A plot study like this grass trial can be used to examine effects of fertilizer, drought, crop species etc. under 'real world' conditions, and can be integrated with laboratory approaches such as pot trials or incubations. *Source:* Sara Vero.

Who Does Fieldwork?

Researchers at almost any stage of their career may undertake in fieldwork, although frequently, the amount of time an individual spends in the field will probably decrease as they move toward a more senior or supervisory position (Fig. 1.6). Fieldwork is an excellent teaching tool for bringing

Fig. 1.6 Fieldwork is an opportunity to learn practical skills and apply lessons learned in the lecture theatre or classroom and to be mentored and trained. *Source:* Jaclyn Fiola.

relevance and "real-world" meaning to processes taught in classroom or laboratory setting, both in the secondary and high school setting, and at the undergraduate and graduate levels. In these cases, fieldwork often consists of tours, expeditions, demonstrations, or very structured experiments under the supervision of an experienced tutor or guide. Maskall and Stokes (2008) reported that although there is little empirical evidence that fieldwork quantitatively improves learning, it is generally viewed with enthusiasm by both students and their teachers. Why is this the case? Perhaps it reflects genuine interest held by those individuals either teaching or seeking to learn about the outdoors, for whom classroom activities, while vital, are not complete on their own. Perhaps it is that the tactile and tangible experiences in a "real-world" setting enhance conceptual knowledge and demonstrate its application. Fieldwork teaches students practical and communication skills, contextual understanding, critical and "big-picture" thinking, and the capacity to manage sometimes challenging tasks. These qualities are immensely valuable, both to the individual and to prospective employers, but may not be truly reflected in standard assessment. Sadly, it seems that fieldwork for pre-university students is declining due to a number of factors, including funding and associated costs, implicit hazards and risks, and the move toward computational research in the environmental sciences. This is also true in postgraduate research and throughout industry and academia, as more powerful computational models are widely and cheaply available (Kirkby, 2004). It should be remembered however, that field research is still an indispensable component of modeling. Direct measurements provide the data by which models are built, calibrated, and tested, thus ensuring accuracy and realism. Field and model approaches should not be considered as completely separate approaches to agricultural and environmental research. Rather, they are tools which can be used in conjunction with one another, to build conceptual understandings and examine hypotheses. I hope that educators reading this guide will consider the great advantages and opportunities offered by fieldwork and will resist the trend to remove it from their curriculums.

Thankfully, many undergraduate students still take courses that are either wholly or in part field based and may conduct individual or group fieldwork projects. At this stage of an individual's education, they are likely to be specializing and honing in on their area of interest. Fieldwork at this stage not only teaches the student but better enables them to learn in the future, by exposing them to challenges, forcing them to apply their existing knowledge, adapt to new situations, and work with other people. At the undergraduate level, field research advances students' knowledge, provides realistic, hands-on learning opportunities, develops critical thinking and problem solving, and communication skills and teamwork (Fig. 1.7). In short, fieldwork helps you *learn to learn*. This is the best lesson of all.

As a masters or doctoral student in any environmental field, you are more than likely to have at least a component of field research. Of course, the type, duration, and goal of fieldwork varies depending on the specific project. As a post-graduate student, *you are a researcher*. While you are under the supervision of an advisor, it is your responsibility to design, conduct, and analyze your own experiment. This will likely change your approach to fieldwork. It is no longer prestructured and prepared by a lecturer or assistant as it is for undergraduate students. You are out there to answer a question. Anticipate that fieldwork may be challenging, both physically and mentally, but if we already knew the answer, there would be no need for your research! Although there are many unknowns, a sound approach to your field research can help you to find that answer (Fig. 1.8).

When we look beyond education, researchers of all ages, career stages, and areas of interest may take to the outdoors to examine hypotheses, develop/test new technologies, monitor responses to change and ground-truth models. Burt and McDonnell (2015) proposed that lateral, novel thinking and constructive debate is constrained by a dearth of fieldwork and the assumption-challenging

Fig. 1.7 In addition to technical skills, fieldwork teaches communication, teamwork, problem solving and planning. These are valuable abilities both for researchers and student who pursue other careers. *Source:* Krista Keels.

Fig. 1.8 A well designed field experiment allows effective data collection and in turn, helps you to examine your hypothesis. Knowing why you are doing this is the first crucial step. *Source:* Rachael Murphy.

experiences that only the field can bring. It seems very likely that this limitation occurs beyond the field of hydrology that they described, and perhaps infects many fields of environmental investigation. Consider this scenario, without the monitoring and examination of diverse or dynamic environments, our understanding of their behaviors is grounded in assumptions made a priori, from potentially very different situations. We may be in error then not because our calculations are intrinsically incorrect or inaccurate, but rather, because they simply do not "fit" the areas we are concerned with.

Why Am I Doing this?

People often question why they are doing fieldwork (sometimes loudly and with profanity) too late. This is often midway through their experiment, with the weather closing in, as they are struggling to collect samples! It is actually the most important question you can ask yourself and is the driver for all of the decisions you will make during the planning process. Asking **"Why?"** will help you identify the appropriate design of your experiment. It is important to remember that the experiment should be designed to test your specific hypothesis (possibly excepting case studies – discussed later). You should not choose to perform any fieldwork without examining whether it can provide *appropriate*, *sufficient*, and *timely* data relating to your hypothesis.

Let us briefly unpack these three qualities. Is the data you intend to collect *appropriate* to your hypothesis? For example, if you are examining nitrogen use efficiency in soybeans, you will probably need to account for nutrient inputs, crop uptake, leaching, and gaseous losses. You will also need meteorological and soil data for context. All of this data is relevant to your hypothesis. Other data may not be appropriate. For example, the traits of your soybean species relating to disease resistance might be relevant to soybean research in general, but if it is not a factor in nitrogen use efficiency, then examination of these traits (which are important themselves) are not appropriate to your study.

Sufficient data means that you have enough measurements to satisfactorily answer your research question. That means enough sites, variables, blocks and/or plots, and replicates. In crop studies, it may mean having multiple growing seasons. There is no real rule of thumb for this. Let the literature guide you and if possible, consult a statistician *before* beginning your experiment. It is not pleasant to realize that more replications are needed once a study has started, and even worse to discover once the field trial is "finished"!

Finally, you need *timely* data. That means that both the collection and analysis of the data is physically achievable for you and your coinvestigators, and ultimately, possible within your project timeframe. This is closely related to having sufficient data, and a balance must be achieved between gathering enough samples and completing your research in good time. If you have embarked on a two-year MS program, there is little point in designing a field experiment involving three years of monitoring! This might seem obvious, but it is very common for fieldwork to overrun projected timescales.

Particularly in the early weeks of a project, it is common for individuals to rush into fieldwork due to enthusiasm and eagerness to collect data – any data! However, if the data is not relevant, the methods or site are inappropriate or the experimental design is weak, the fieldwork often results in much wasted time and resources and flaws that are revealed during the peer-review process. There may be lessons learned in this process, but that time would be better spent in scrutinizing "why do this?" and in planning, so that your work can succeed from the beginning. As an example, I once was asked to give advice to a PhD student in relation to a sampling campaign he had embarked on. When I asked why he had chosen to collect samples from a certain location near his campus, he responded "Well it was convenient, and I needed to get started." However, in his rush for data, he had chosen a site that was completely unsuitable to the method of analysis he was developing, and so had wasted months of valuable time.

Some whys:

- To obtain data that will allow you to examine a hypothesis or question
- To provide contextual information
- To examine "real-world" case studies
- To gain skills and experience
- To test the performance of devices, processes, or tools

For graduate students there are two main concurrent goals. First, to explore and hopefully fill a knowledge gap in their particular discipline, by conducting a series of related experiments culminating in a thesis. The other goal is to develop skills that they can apply to other research projects or to industry. This goal is by no means less important than the scientific objective. Fieldwork is a fantastic opportunity to develop practical skills, problem solving, logistical skills, and teamwork. It is often relatively simple to get a sensor working in the lab when you have tools at hand, good lighting, and someone to ask for help. Can you do it on a rainy day, far from your office when you can't find your screwdriver? From an employer's perspective, the exact detail of your (no doubt important and complex) past research may or may not be interesting or applicable to the role you are interviewing for, but your ability to learn and apply practical skills will help you to be useful and effective in any situation.

Before embarking on a campaign of fieldwork, ask yourself these questions:

1) What am I trying to examine?

   This is the first and most important question to ask yourself. It must be related to your hypothesis or question. Your fieldwork must do one (or both) of the following:

   A) Provide information that either proves or disproves your hypothesis. For example, your hypothesis might be "Soil compaction reduces the yield of perennial ryegrass over three years." Your fieldwork must then measure differences in the yield of perennial ryegrass under both compacted and non-compacted conditions. Critically, it must be measured over three years. A two-year study won't answer your hypothesis!

   B) Provide contextual information that helps explain your results or their implications. For example, let's say you conduct an incubation study in the laboratory to measure denitrification in soil at various temperatures. In that scenario, you can impose whatever temperatures you like. What temperatures are actually encountered in a real soil, outdoors and exposed to the weather? So, you might conduct fieldwork in which you install sensors to measure temperature over time. This provides context for your laboratory study and helps you to examine its relevance in the discussion section of your paper or thesis.

2) What treatments do I need to examine my hypothesis?

   A **treatment** is the condition, practice, or manipulation that you apply to your field site. This will depend entirely on your hypothesis, and in many field studies, multiple treatments may be applied either in isolation or in combination with one another. In surveys or case studies, there may not be any treatments at all, since you are measuring the state of a person, place, animal, or thing, or are documenting a particular event or situation. When a treatment is applied, there should always be a control that is not modified, against which you can compare your results.

3) What do I need to measure?

Again, this comes down to your hypothesis. You need to measure the variables that you expect might be altered by your treatment. If you are examining the effects of light pollution in urban areas on the behavior of a certain bird species, then you might want to measure how frequently the birds eat, sleep, mate, or sing. However, you must also quantify the treatment. In this case, how many hours of light are the birds subject to, relative to a non-polluted situation?

Collecting supplementary information during your fieldwork is also useful because it either helps explain what you observed or it may help your reader to evaluate how applicable or transferable your results are to their own situation (Fig. 1.9). Some examples are location (latitude, longitude, elevation, country) or weather variables (precipitation, temperature, humidity). There are, of course, many other details that are relevant in different fields of study, so think carefully before launching your field campaign. It is usually far easier to collect data at the time rather than returning for further data collection when a reviewer has asked for it! Look at the literature on related studies as a guide. For example, if everyone conducting a river survey typically describes whether their study area is a first, second, or third order stream, this is a good indicator to you that this is highly relevant information. As you become more familiar with your field of study identifying what contextual measurements you should take will become more obvious.

4) How should I take measurements?

There are often several methods or tools available for measuring a certain parameter. For example, soil hydraulic conductivity can be measured using a double-ring infiltrometer, a transducer infiltrometer, an Amoozemeter, a falling head test, a constant head test, and others. The differences between these devices might seem subtle or minor until you become familiar with them, but might be vital when it comes to interpreting your results. There is often an element of availability that needs to be considered here. What tools and facilities can you access? Can your

Fig. 1.9 Weather information is one category of supplementary data that can be helpful in interpreting or contextualising the results of your field experiment, although it can be useful in it's own right also. *Source:* Sara Vero.

university or research institute provide training, or can you buy or rent equipment? Can you outsource analysis for specific tests? How time-consuming and how expensive are the various options?

Most importantly, be thorough in your literature review. Consider what methods are used in related research and why. Don't be afraid to contact the authors of the papers which you are referencing. They will often be able to explain exactly why they used particular methodologies and to offer advice.

5) What characteristics should my site have?

This is a multifaceted question so let us break it down into manageable parts:

A) **Does the site have the characteristics which will allow you to test your hypothesis?** In other words, if your hypothesis is that slope aspect influences the rate of snowmelt in alpine mountain ranges then your field sites must (i) be alpine mountains, (ii) have snow-cover for a given period, and (iii) exhibit a range of different slope aspects that you can compare. This might seem obvious, but it is remarkably easy to choose sites based on ease of access, familiarity, or other generally positive traits that are actually poorly suited to the hypothesis in question. When choosing sites that are intended to be representative of a particular environment or situation you must think carefully about what the defining characteristics are and list them. Then, you can review potential sites objectively.

B) **Does the site have any characteristics that might unduly influence or confound the examination of your hypothesis?** So, in your alpine snowmelt study, is one of your potential sites heavily forested while the others are relatively bare? If so, then this site is probably not suitable for inclusion in your study as there are other factors that might overly influence your results.

C) **Can you access the site?** Even if you identify a site that is ideal on paper (it has all the characteristics of the scenario you want to study and it is suitable for application of your treatments), there are also logistical concerns. How far away is it? How long will it take you to get there and back? Is there electricity, water, or other facilities you might need? Is the landowner willing to grant you access? How close can a road get you to the site and are you capable of transporting your equipment across fields, rivers or hills? Is it safe? Are there any potentially dangerous animals? Be realistic in evaluating these issues.

6) How long will fieldwork take?

Fieldwork can be more time-consuming than expected on paper and as you are subject to the environment and the unexpected (loss of tools, breakdowns, bad traffic, poor conditions, etc.), it is vital that you schedule extra time for these possibilities. Your planning might look something like this:

Travel time + Setup + Treatments + Measurements + Rest + 'The Unexpected'

Travel time can be estimated with reasonable accuracy from route planning tools such as Google maps; however, you should allow extra time for traffic. If you are bringing heavy equipment such as a trailer, you may also be slower than otherwise expected. You can't really attach a precise number of hours to the unexpected events that can occur during fieldwork. Rather, schedule some spare time to allow for unplanned circumstances. You can help minimize these by preparing thoroughly, using checklists, and practicing with your equipment in advance. Setup and application of treatments and measurements can be estimated in advance if you conduct a trial run. The more familiar and practiced you are with the tools and techniques you will use, the more efficient you will be. Never try a technique or tool out for the first time in the field. Rest is also critical. Don't expect that you will be able to squeeze in an extra few measurements at the expense of a lunch or coffee break. This may be unavoidable sometimes, but if you are engaged in a prolonged fieldwork campaign, it will wear you down and ultimately make you less effective. If you are driving to your field site, this becomes even more important. Tiredness is sadly a common cause of road accidents. If you feel yourself becoming sleepy while at the wheel (considering sometimes, long distances to field sites and the strenuous nature of the work this is not unusual), pull over where it is safe to do so, take a nap, eat, drink a coffee or energy drink, or if necessary and possible to do so, rest overnight. If you are traveling with someone, you may want to share the driving. Fieldwork is no excuse for being irresponsible in this regard. Unsafe driving puts both yourself and others at risk. This is discussed in detail later on.

You should also consider the importance of seasons or years. Many environmental and agricultural research projects need multiple seasons or years to allow full examination of a treatment. While this might be fine for permanent researchers, if you are a PhD student you typically have a very finite length of time in which you must gather, analyze, and write-up your data. If your PhD program is three years long and allowing time for design and setup prior to the experiment, and time for thesis preparation subsequently, achieving two years of field data may be challenging. However, multiple seasons or years usually add greater reliability to your findings. The fewer seasons of data you report, the greater the likelihood that your data and interpretations will be influenced by factors such as weather specific to that year. It is up to you and your advisors to figure out the optimal approach here. This is something that you should consider prior to starting your field campaign. Remember, you can usually spend as much or as little time in the laboratory as you find necessary. When it comes to field seasons however, we are all at the mercy of time!

7) What statistical structure will I use?

It is good practice to consider your statistical approach or even to seek the advice of a statistician when initially planning your fieldwork. If you are conducting a plot or other replicated study, this will help you to select the optimum number of replicates. If you are conducting a case study, statistical approaches may not be easily implementable as the variables in this type of research are not always strictly controlled. In such cases, a thorough and well-developed discussion of your results is particularly important.

8) How much will this experiment cost?

It is all too easy to design large-scale, comprehensive, and high-tech field studies. In reality, costs must be accounted for as this will constrain your plans. Some general costs you need to consider include:

Staff	Consumables	Licenses/visas/permits
Vehicles/fuel	Accommodation	Hardware/Software
Equipment	Contractors	Analyses

Field studies can incur large costs, and it is important to be realistic in estimating these at the outset. If your intended study includes a field component over many years you should firmly establish funding in advance. Many research institutes will have a financial department that can help you to calculate your budget, but it is up to you to determine the costs of equipment, consumables, and analyses that you will require. Keep in mind that you may not have the skills, equipment, or authority to perform certain tasks and may need to hire contractors. Your research institute may have established relationships with these contractors or else you may need to shop around and obtain quotes for the intended work. Just as additional time is sometimes required, there can also be unexpected costs, such as replacements for equipment, additional samples to be processed, etc. Leave yourself some extra budget, if possible, to cover these eventualities.

9) Do I have the equipment?

Equipment includes tools, safety gear, appropriate clothing, machinery, and sampling or measurement devices that you will need during your field study. Don't assume your research center will have exactly what you need, and be very wary if someone tells you "Oh, we have one of those in the shed..." Stored equipment should always be carefully examined and tested to see that it is functioning correctly and can actually supply the results you need. Sometimes older equipment can be repaired or refurbished if it has only minor damage. Often simply replacing dry and cracked tubing, O-rings, and other rubberized components is all that is required as these parts tend to degrade in storage. Check whether such equipment has been surpassed by more modern devices. These may offer superior measurements, greater ease of use, or will be more understandable to the modern readers of your studies.

If you are purchasing new equipment or if you have several different devices at your disposal that measure the same variable, it is a good idea to speak to an expert who thoroughly understands the various approaches and can advise you as to which would best serve your purpose. It's vital to do your background research also. Typically, newer methods or devices will be tested against the older, more established approaches. Look for peer-reviewed research on your intended methods and consider emailing the corresponding author if you have further questions.

If equipment does become damaged, think carefully before attempting to repair it. Some items can be repaired relatively easily, but don't assume that you have the skills to do so unless you have been appropriately trained. This is particularly true for equipment which has mechanical or electronic components. Equipment is often purchased with warranties and service agreements. Check that you will not invalidate these by tampering with the device! For "simple" repairs or maintenance, such as replacing hoses or cleaning sensors, consult the manuals and use the correct tools. Much hardship and frustration can be saved by reading the instructions before you take things apart!

If your research institute has equipment that is available to many different people, be sure to check with whoever is responsible for its storage or maintenance before you take it. There may be a schedule or roster for its use and taking equipment without following the proper routine can infringe on other people's work. Remember, your project is no more or less important than theirs! Always return used equipment in a timely manner and in good condition.

Clean tools after you have used them and put them in their proper place (Fig. 1.10). Be respectful.

Fig. 1.10 Always clean and store equipment properly after using it. Don't leave it in poor condition for the next person. *Source:* Sara Vero.

10) Do I have the skills?

It will probably be obvious that you will need the scientific skills related to applying your treatment, and correctly collecting, processing, storing, and analyzing your samples. These skills will often be learned at your university, through on-the-job training, by reviewing standard operating procedures (SOPs), or by seeking specific training via workshops, courses, or online. Think about what other skills you might need that are not "scientific" in nature but will be required. For example, can you read a map, mark out plots accurately, drive off-road, pull a trailer, use essential machinery, etc. The list is endless! Some tasks may not only require that you *can* do them, but also that you have training, licenses, and/or approval from your institute or under the law. Driving skills come under this category.

Where you are lacking certain skills, you have three options.

1) **Learn** – The major advantage here is that you can accrue new abilities that will help you in your future research or work and this should always be a priority throughout your career, particularly in the early stages. However, in some instances it may be too time-consuming or inefficient from a cost perspective and it would be better to seek assistance (Fig. 1.11).

2) **Teamwork** – Much of the best research is done in teams and collaboration is one of the most important abilities a researcher must establish (Fig. 1.12). Fieldwork is an excellent way to learn this and allows more ambitious and comprehensive studies than are usually possible by one or two individuals in isolation. We will discuss later how to organize your team.

3) **Contractors** – Some tasks are better outsourced to people or agencies who specialize in this area. This could be because it is a once-off task and it may not be worth your time developing, or it could be that the task is so specialized that it requires an expert. Machinery

Fig. 1.11 Collaboration in the field is a great opportunity to learn new skills from experienced researchers. *Source:* Brandon Forsythe.

Fig. 1.12 Sometimes additional help may be necessary. This team is working together to handle large, awkward equipment. *Source:* Katie O'Reilly.

use also is often best outsourced. Hiring a crane operator, for example, will likely be preferable than learning to use one and then renting the crane itself! Liaising with contractors is essential to ensure that everyone knows their task, is in agreement regarding the price and will be at the right place at the right time.

11) What training do I need?

Methods and equipment – You should be confident that you understand all of the methods and equipment you intend to use both in theory and in practice. While reading the manual is an important starting point, it is not always enough! Remember the six Ps–Proper Planning and Practice Prevent Poor Performance.

Health and safety – Check your institutional policies before commencing either field or labwork. There are often mandatory requirements regarding manual handling training, basic first aid, biosecurity, or other health and safety protocols. You may also consider specific training that might be relevant to your intended research (Fig. 1.13). For example, if you are researching infection rates of toxoplasmosis in urban feral cats you might need training in animal handling, zoonosis, and ethics relating to research with animal subjects. Research institutes typically are very supportive of training, especially on health and safety issues.

Training takes time. This should be accounted for in addition to the fieldwork itself as it should be completed before you venture into the outdoors. Although fieldwork is certainly an opportunity to learn and develop new skills, the first time you try a new technique, tool, or idea should never be during the "live event," when you need everything to run smoothly.

Don't underestimate the value of informal practice. Set aside some time to trial run your equipment and techniques. This will ultimately save time in the field, prevent damage to equipment, allow effective measurements, and prevent a great deal of frustration.

Fig. 1.13 You may need training in safety protocols. For example, the researchers investigating prairie burning in this photo from Konza Prairie, Kansas, have training in fire safety and emergency response. *Source:* Jesse Nippert.

12) What assistance do I need?

Broadly speaking, there are two factors to consider when determining how many people you need on your team and who they should be.

Skills – Often you will require specific skills on your team that include experience, specialist training and equipment, expert knowledge or all of these factors. In these situations, it may be best to recruit someone to your team who has these skills. For example, if you are primarily a hydrochemist studying pesticide transport to groundwater and you need an accurate characterization of the soil profile, you probably should seek the assistance of a pedologist with both training and experience. This is a great opportunity to learn from these individuals.

Labor – Many hands make light work and furthermore, some jobs are simply not safe and/or possible for a single individual. An example from my own research is a series of river surveys during low flow. In that study, I needed approximately 50 water and sediment samples from across entire river and tributary networks, and they needed to be taken in the space of roughly 4 h. Alone, that would have been completely impossible. However, the sampling methodology was relatively simple and needed only brief training. From my research center, five people lent their assistance each day. We met each morning, I assigned everyone a specific stretch of river and the team regrouped once their samples had been collected. The density and tight timing of sampling could never have been achieved with fewer people, no matter how skilled or motivated.

Students and researchers are generally helpful and enthusiastic. Most researchers can probably relate plenty of stories of teamwork generously and freely volunteered between friends. I can list a dozen friends and colleagues who readily contributed their time and effort during my PhD alone. Collaboration is built on reciprocation. Depending on an individual's contribution they may merit inclusion as a co-author, a collaborator, in acknowledgments, or by some other recognition of their input. This can only be determined on a case-by-case basis and should be reviewed in light of your institutional policies. When possible and appropriate, try to contribute your skills and assistance to others in return.

Outside of collaborators, you may need to hire assistance. This is especially true where skills or equipment are outside of your expertise. For example, if you need electricity to feed a river bankside analyzer this should not be considered to be an opportunity to become an electrician! This needs skill, experience and tools both to do the job correctly and to safeguard your health and that of your team. Hire a registered electrician!

13) Do I need permission from landowners, local authorities, managers, etc.?

Some fieldwork may be conducted on study sites that are owned by research institutes or on which the institutes have agreed access to. In this case, you most likely need approval from senior managers or officials within your institute. It is best practice to also communicate with whatever staff are responsible for the day to day management (e.g., the farm manager on an agricultural research station). This will prevent either your interruption of their work, and vice versa. You should inform them of where exactly you will be, for how long, what you intend to do and if there is anything you need them to do or avoid doing. For example, if you are operating a gas flux tower on a research farm you may want the staff to refrain from allowing livestock into that area of the site unless additional fencing has been installed.

When operating outside of a research facility there are some other concerns that you should consider. If you want to conduct research on privately owned land you will need explicit permission from the land owner. It is best practice to obtain this in writing. The same rule applies if you need to cross someone's property to access a site, even if they themselves do not own that site. Remember, your work does not take precedence over a right to property or privacy and is *never* an excuse for trespassing. It is important to be clear about what you intend to do. Are you digging a soil pit? Taking water samples from their stream or well? It will help your work immensely if you maintain good relationships with any property owners that you interact with. This is particularly the case where you will be onsite on many occasions. The best way to do this is by being respectful of their time, property and privacy, and by communicating appropriately.

Privacy is an important consideration. It is very common to obscure the identification of on-farm sites in publication for confidentiality, as landowners may not wish for their home or business location to be disclosed. This is an entirely reasonable request and should always be respected. In Europe, this is the law (General Data Protection Regulation 2016/679). Equally, where farmers, community groups or other stakeholders have contributed to your research (by supplying sites, providing data, completing surveys, etc.) their contributions should be acknowledged in talks or presentations that might arise from your results. Farmers may request to see your results prior to publication, sometimes out of concern for potential repercussions but often to improve their own practices or better understand their land. Local farm discussion groups, fishing and hunting clubs, outdoor enthusiasts, and community associations frequently welcome researchers operating in their areas to give talks and explain their findings. This is helpful, both in disseminating your work and in building positive relationships with the community. It also enables them to enact changes in their practices to improve their environment, or to simply better understand it. Good relationships with your hosts are crucial, and good channels of communication are the foundation of building trust.

Communication is particularly vital if there is any element of hazard or risk involved. As an example, I was once installing a soil sensor array, which involved a >2 m deep pit. The land owner was fully aware and approved of the project. However, at nightfall on the first day we had not finished the installation and the pit was still open. I rang the land owner to inform him and he had no concerns. But your responsibility may go further than that. During the day I had noticed a house nearby the field with three young children playing in the garden. While you or I may quickly recognize the danger of an open pit, a 6-yr old probably would not, and may find it all too interesting! I set up warning signs around the area and then politely introduced myself to the family. I presented identification, explained what I was doing on the nearby farm and explained that the pit would be open overnight but that we would return to complete the installation and backfill it in the morning. As a result, I could rest easily that evening knowing that I had alerted the young family to the potential hazard and taken steps to mitigate the risk.

You may be conducting fieldwork in a public or common area. It is best to consult your local authorities (e.g., county council, park ranger, etc.) if this is the case. Your research institute may have ongoing agreements or relationships that can help you in this, but if you are proposing entirely new research at a site that has not previously been used by you or your colleagues it is best to provide a clear and comprehensive description of what you are planning to do and any related considerations. Again, written permission is important. Don't assume that it is permissible to do whatever you have planned in a public area. You may need to file permits, seek derogation, or to provide advance public notice, particularly if your planned fieldwork is invasive or disruptive to other people, or is perceived as being so.

Maybe you'll be fortunate enough to do research in an exciting or dramatic location; the Antarctic, rainforests, areas of conservation. Or perhaps in areas that are particularly dangerous, such as radioactive or contaminated zones. You may need permission or simply to inform the department or agency responsible for those areas. Similarly, if you are traveling abroad for research make sure you have the appropriate Visa. This will in many cases not be the same as a holiday or work visa that you may be familiar with. If you encounter any difficulties be prepared to contact your embassy.

References

Burt, T.P. and McDonnell, J.J. (2015). Whither field hydrology? The need for discovery science and outrageous hydrological hypotheses. *Water Resources Research* 51, 5919–5928.

Kirkby, M.J. (ed.) (2004). *Geomorphology: Critical Concepts in Geography*. Volume II. London: Routledge: Hillslope Geomorphology.

Maskall, J. and Stokes, J. (2008). *Designing Effective Fieldwork for the Environmental and Natural Sciences. GEES Subject Centre Learning and Teaching Guide*. London: GEES Subject Centre.

Read, D.J. (2003) Towards Ecological relevance — Progress and pitfalls in the path towards an understanding of mycorrhizal functions in nature. In: van der Heijden M.G.A., I.R. Sanders (eds.) *Mycorrhizal Ecology*. Ecological Studies (Analysis and Synthesis), 157. Berlin, Heidelberg: Springer.

2

Types of Fieldwork

Experimental Design

Plot Experiment

A plot experiment consists of treatments and controls arranged in experimental units of specified area, measuring the effects of those treatments (usually over time) and applying statistical analyses (Figs. 2.1–2.3). Experimental units are the individual plots or pots to which experimental treatments and their replication are independently applied (Fig. 2.4). Plot experiments are very common in agricultural research such as crop and fertilizer trials, studies of vehicle effects, and leaching and runoff studies. A major advantage of this experimental approach is that it allows multiple treatments and combinations thereof to be examined at the same time. Plot experiments are well suited to statistical analyses provided sufficient replication of plots is achieved. Individual plots can receive a variety of different measurements that can be repeated across time. For example, a plot trial used to investigate grass yield under different soil phosphorus indices could include measurements of crop (total yield, dry matter, and nutrient concentration), soil

Fig. 2.1 Plot and row experiments at a crops research station. *Source:* Colum Kennedy, Teagasc.

Fieldwork Ready: An Introductory Guide to Field Research for Agriculture, Environment, and Soil Scientists, First Edition. Sara E. Vero.
© 2021 American Society of Agronomy, Inc., Crop Science Society of America, Inc., and Soil Science Society of America, Inc. Published 2021 by John Wiley & Sons, Inc.
doi:10.2134/fieldwork.c2

Fig. 2.2 Signs like these can be helpful in identifying which plots receive certain treatments. *Source:* Jaclyn Fiola.

Fig. 2.3 Care should be taken when harvesting field trials to accurately record yields without damaging the plot. Make sure that your mower has appropriate tyre pressure and that the blades are set to the correct height. *Source:* Sara Vero.

(phosphorus concentration, soil moisture), and water (nutrient concentrations in pore water in the root zone) parameters. The variety of measurements would allow a comprehensive understanding to be developed, while repetition over time may support statistical analyses. It is best to have multiple plots receiving identical treatments as natural heterogeneity of soil and landscape, even over relatively small areas, can influence measurements. Another consideration in plot studies is "edge effect." This occurs where plots located at the edge of a row or block are subject to slightly different conditions than those in the center and may be less buffered from conditions in the rest of the field.

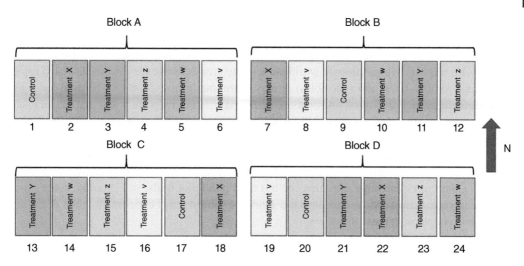

Fig. 2.4 Example of a field plot layout including four blocks, each with five treatments and one control. Treatments within each block are randomized. *Source:* Sara Vero.

The size of plots can vary immensely, from a couple of square meters to "strip" plots, encompassing entire crop rows or even fields or parcels measuring several hectares. Smaller plots allow more treatments to be applied but may suffer from greater edge effects and the measurements within each plot will be more subject to site-specific variables. Larger plots or strips may smooth out these influences but are logistically demanding. Field research is often limited by available area (particularly if you are being hosted by a participating farmer rather than a research farm owned by your institution or university), and large-scale trials are more expensive to establish and maintain. Irrespective of the size of your plots it is important to impose adequate spacing or barriers between them so that the treatments applied to each plot do not affect those adjacent to it. For example, if you are examining the efficacy of an herbicide on mixed species swards, airborne drift of the chemical could lead inadvertently contaminate other plots. Be sure to account for spacing in your initial design and layout of your plots.

Another consideration in determining plot size is the nature of the treatment. Some treatments are well suited to plots measuring just a few square meters, for example, seed mixes or herbicide treatments. Others can only be implemented at a row or field scale to be representative of farm practice, because of the size of machinery involved or due to the size of plant species being examined. Treatments requiring this larger scale approach might include cultivation practices, investigation of subsurface drainage systems, or testing of irrigation systems (among many others).

Plot experiments are well suited to randomization (e.g., complete or incomplete randomized block design, split plot arrangements, multifactorial design, etc.). Some excellent and detailed aids to selecting the appropriate treatment design are provided by Oehlert (2010), Cassler (2014), and Glaz and Yeater (2017).

A consideration when running plot trials (or row trials for grain crops) are edge effects. Edge effects manifest as differences in growth pattern near the perimeter of plot or row trails, compared with at the center. There are two types of edge effect: border effect and neighbor effect (Langton, 1990). Border effect occurs because plants growing along the perimeter experience slightly different conditions than those in the center. For example, they may be more exposed to wind or inclement weather. Neighbor effect occurs when a plot is influenced by neighboring treatments, plots, or untreated ground outside the plot margin. Examples of this include drift of herbicide treatment or interaction of plant roots below the surface. Edge effects are more pronounced in small plots, so a common preventative measure is simply to use larger plots. You should consider whether this is possible with respect to the additional workload and the area available for your trial. Guard rows (grain crops, shrubs, trees, and vegetables) or edge trimming (grasses) can also be used to avoid edge effects. In using guard rows, one or more additional rows of crop are planted at the edge of each plot or row (Fig. 2.5a). These are discarded before sampling the main plot itself. Similarly, edge trimming means that a predetermined area of the plot is cut and discarded, while only measurements taken from the center of the plot are used for analysis (Fig. 2.5b) (Peterson, 1994). Alleys (gaps between plots) or temporary screens can be used to avoid treatment drift between neighboring plots. Guard rows and edge trimming are not used in all plot trials; you should consider the literature in your subject area and the specific details of your planned experiment when deciding if and how to implement these measures.

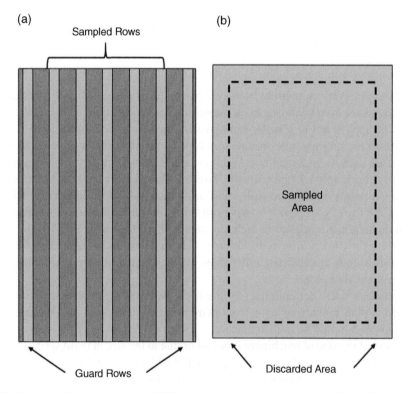

Fig. 2.5 (a) Guard rows in row crop trials. (b) Discard area and sampled area in a plot trial.

Survey

In field surveys, the researcher collects data from across an area with a view to defining spatial characteristics or patterns (Figs. 2.6 and 2.7). Surveys may be repeated to determine trends over time. Depending on the size and nature of the study area, surveying can be time-hungry and physically challenging. Frequently (although not always), survey data can be expressed using maps, which is engaging for the audience or reader if done well.

Fig. 2.6   Researchers embarking on a survey in the mountains. *Source:* Jaclyn Fiola.

Fig. 2.7   Soil surveys are just one form of field work used to develop and validate soil maps. *Source:* Sara Vero.

Survey types in environmental science include:

- Soil mapping
- River chemistry snapshots
- Groundwater well dipping (for determining watertable depth)
- Species characterization (plant or animal)
- Topographic mapping
- Geological survey

Although it is not strictly a biophysical method, surveys of farm management such as fertilizer rates, timing of cultivation, and stocking rates have an important role in agricultural and environmental research. These factors may influence your observations and are crucial to interpretation of results. Recording farm practices requires the involvement of the farmer or farmers on whose land the research is conducted or who live within a study catchment or watershed. For catchment-scale studies (in agricultural landscapes), it is desirable to have a representative proportion of farmers participating in record-keeping. These records can be demanding, both for the researcher and the participants so it is helpful to be clear at the initiation of the study what each party is expected to record and how they are to do it. Hard-copy or online diaries help the farmer to keep track of their activities, particularly those details which you are interested in.

Case Study

A case study is a detailed investigation of a specific scenario, location, or occurrence (Fig. 2.8). Case studies offer insight into real-world situations and allow the validity and applicability of hypotheses and theories to be evaluated in-vivo. Extrapolation of conclusions from case studies to other locations and situations should be conducted with care, since experimental conditions can often vary from other situations. In other words, you must determine whether the two situations are sufficiently similar. Furthermore, a case study is essentially a snapshot; your observations will be influenced by a variety of factors affecting the case at that time. On the positive site, case studies are essential for translating scientific knowledge into practice and critically, can open up new

Fig. 2.8 The researchers in this photo are collecting soil, water and ecological information to form a comprehensive characterisation of their case study. *Source:* Jaclyn Fiola.

avenues of research, reveal knowledge gaps and provide an insight into the environmental context which is sometimes lacking from studies confined to the laboratory or desktop. Aside from research, case studies are an excellent teaching tool, allowing students to examine and observe the reality of their fields of study.

Case studies can be representative in nature. In other words, the sites and scenarios examined are selected based on how representative they are of a broader environment and the results and observations derived may be generalized or applied to this wider context. For example, a case study of the changes to fertilizer management on a family farm after removal of milk quotas in Ireland could be used to illustrate changes to the sector as a whole. Representative case studies can be partnered with surveys or controlled studies to demonstrate how their findings translate to reality. Where funding and resources are available, a number of case studies can be collected to provide a more comprehensive representation.

Conversely, case studies can be used to document and demonstrate outliers, in which an unusual, anomalous, or atypical event or situation occurs. These case studies are vital for identifying knowledge gaps.

Fig. 2.9 Equipment like this eddy covariance tower can be used to monitor atmospheric conditions including temperature, wind speed and direction, carbon dioxide, methane and other gas fluxes. *Source:* Rachael Murphy.

Monitoring

Environmental monitoring has a long history and has been a cornerstone in the development of the fields of hydrology, contaminant transport, air quality, ecology, and more. Monitoring is the ongoing observation, recording, and characterization of a site, process, or variable (Figs. 2.9 and 2.10). This type of research allows trends (directional patterns) to be detected and evaluated and typically these studies gain power and reliability as they become more prolonged and as the frequency of measurements increases. For agricultural and environmental field research, single seasons of observation can be influenced by the unique weather conditions or distinct events occurring within that period. This can constrain the conclusions which may be drawn from that study and may limit or influence how they can be applied to a broader context or to other situations. Furthermore, many environmental changes are relatively slow, having innate time lags due to geographic scale, rate of water movement, length of breeding cycles, and a host of other factors. Monitoring over sufficiently long periods is therefore essential to evaluate both environmental impacts and the processes which influence them (Fig. 2.11). Field monitoring is often used to determine compliance with environmental legislation and to supply evidence for the enforcement of those laws.

Fig. 2.10 This phenocam at Konza Prairie, Kansas, provides automated recording of plant canopies and is part of a network across the United States and Canada. Collections of case studies using identical methods can allow greater conclusions to be drawn. *Source:* Sara Vero

Fig. 2.11 Monitoring infrastructure such as the weather station at this farm research platform in the United Kingdom can be powered by solar panels. These reduce the need for frequent battery replacement and are capable of supporting relatively demanding equipment. Also seen here is a livestock-proof fence - essential for preventing damage to the weather station. *Source:* Sara Vero.

Hydrology is perhaps notable for utilizing long-term monitoring studies, some spanning over multiple decades. Perhaps this stems from our current and historic reliance on watercourses for abstraction, transport, and fishing and conversely, the potentially catastrophic threat of floods. The River Thames in London, U.K. is an example of long-term monitoring and provides the longest record of water chemistry in the world. Monthly nitrate concentrations have been recorded for over 140 years, starting in 1868, accompanied by weather records for the same period and discharge since 1884. This remarkable record was investigated and documented by Howden et al. (2010), but the initiation of the monitoring was done by drinking water treatment works supplying the city of London. The engineers who established this likely had no idea that the records they began would provide insight into the environmental consequences of population increases throughout the 20th century, the advent of chemical fertilizers, World War I and II, land-use changes, the establishment of the European Union and the water and agricultural laws brought in thereafter. While the extensive record allows each of these historical events to be examined, it also informs the design of other monitoring endeavors. For example, by evaluating the rate of hydrochemical change, the authors of that study determined that studies of shorter than 15 years would be vulnerable to error if lacking appropriate historical context. The design of legislation also depends on this evidence to guide expectations of environmental responses, which may not correspond to governance or election cycles. The definition of "long term" research varies between disciplines; however, some general consensus appears to be around 10–15 years. Lindenmayer and Likens (2010) proposed a 10-yr threshold for ecological monitoring.

While no strict rule or agreed convention exists, short-term monitoring may lend itself more to case studies, while increasing length and frequency of monitoring allows application of more statistical analyses.

Monitoring studies can take different approaches including (but not limited to):

- Repeated physical sampling of water, soil, or vegetation for analysis at the laboratory. This samples can be obtained directly by a researcher in the field, or by automated samplers.
- Use of sensors at appropriate temporal resolution for measurements such as temperature, river discharge, turbidity, eddy covariance, etc. Sensors often facilitate high-temporal resolution monitoring up to sub-hourly frequency.
- In situ measurements (often coupled with electronic sensors and validated against laboratory samples). Monitoring at river outlets may take this approach, in which bankside devices automatically extract samples from the watercourse and analyze them on location for nitrogen and phosphorus.
- Observational monitoring may be used for wildlife studies. This can take the form of GPS tagging of birds, fish, or animals, the use of catch-and-release traps, or of field cameras to observe activity and behavior.

Monitoring can be expensive including the initial outlay for establishment of the experiment, its ongoing maintenance and its high demand for consumables. Large monitoring endeavors often require dedicated staff for maintenance of equipment. However, these challenges can be overcome and increasingly the value of monitoring studies is appreciated, particularly for providing baseline or background data for other research.

There are a number of groups and consortiums comprising discrete monitoring projects who collaborate across sites or adhere to agreed standards, measurements, and protocols. These programs might focus on one particular field of research or may take an integrative approach incorporating many distinct fields. As an example, the Long-Term Ecological Research (LTER) Network includes 26 independent research sites funded by the U.S. National Science Foundation (NSF) since 1980. These sites represent a breadth of ecosystems including tallgrass prairie (Konza, Kansas), the Antarctic (Palmer Station, Anvers Island), and marine (California Current). The Long-Term Agroecosystem Research (LTAR) network created by the USDA Agricultural Research Service (USDA-ARS) similarly co-ordinates 18 independent sites across the contiguous United States. The goal of LTAR is to investigate and develop strategies for the sustainability of agricultural production under the three pillars of productivity, environment, and rural prosperity.

These distinct sites follow a consistent approach to data collection across the entire network, although specific measured variables are selected as appropriate to each site (for example, depth of permafrost is measured at the Artic site but would be irrelevant for the urban biome in Arizona). These measurements allow conceptual understanding of these ecosystems, development of ecological, hydrologic, and biogeochemical models, and collaborative investigations across contrasting environments. Ambitious research programs such as these support long-term monitoring over multiple decades from which projections into the future can be modeled and act as a benchmark against which comparable sites can be evaluated. Furthermore, they provide well-characterized, representative, and secure facilities in which controlled and replicated experiments can be executed. An example of this is the watershed-scale tallgrass prairie experiment that has been conducted at the Konza Prairie LTER site in Kansas. In that experiment, 60 hydrologic watersheds have been subject to treatments including bison and cattle grazing, and five burn frequencies (annual, 2-yr, 4-yr, 20-yr, and >20-yr intervals, in addition to no-burning) since the establishment of the facility in 1972 (it later became one of the six founding LTER sites in 1980). These treatments have provided insights into the implications of management for the remaining US prairie grasslands and species.

Sampling Design

Sampling Patterns

Sampling points may be either controlled or randomized (Fig. 2.12). A stream ecological survey for example (particularly if repeated across multiple seasons or years) will often be conducted at pre-determined points along the watercourse. These may be selected based on multiple criteria for example, hydromorphology, distance from the outlet, proximity to a nutrient point source, access (a practical consideration), or any other aspect identified by the research as being of importance. It is vital that if predetermined sampling points (Fig. 2.13) are used in any type of field survey that the researcher records the reasons for their selection, and when publishing, can explain to the reviewer and reader why they made their decision (Fig. 2.14).

Fig. 2.12   These researchers are soil sampling in Alaska. There are a variety of sampling patterns which may be used and you should consider which approach is most appropriate before embarking on a sampling campaign. *Source:* Jaclyn Fiola.

Fig. 2.13   Coloured flags are used here to mark where the researcher is identifying plant species composition using the Daubenmire method. This experiment is part of the Nutrient Network co-operative, in which consistent sampling methods and patterns are applied at independent sites. *Source:* Jesse Nippert.

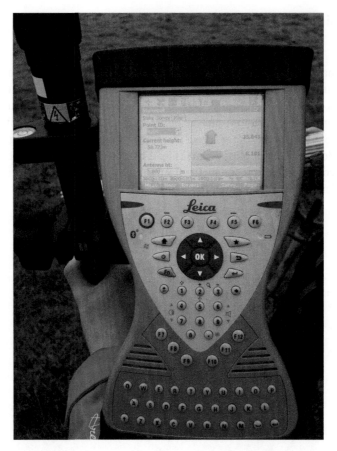

Fig. 2.14 GPS devices can be used to accurately identify sampling locations. These are discussed in the Preparation section. *Source:* Sara Vero.

A variation on this is a semi-randomized approach. For example, where topsoil is sampled for nutrient concentrations, a fixed pattern is walked across the field and an auger is used to take samples to a prescribed depth at regular intervals. The general pattern is pre-arranged, (often a W-pattern (Fig. 2.15a)), but the specific location of each sample is random and designed to incorporate the variability seen across the entire field once the samples are aggregated.

Transect sampling is often used in ecological studies, in which measurements are made at regular intervals along a line through the sampling area (Fig. 2.15b). However, applications do exist in other disciplines such as boring of groundwater wells along hillslopes, or excavation of soil pits along a "catena" (a sequence of related but distinct soil profiles along a slope).

Grid patterns (Fig. 2.15c) allow an even distribution of samples across an area and are often used in precision agriculture to identify variations in soil moisture, nutrients, crop yield, etc. at subfield scale. Detailed discussion of the methods and applications of precision agriculture approaches are available in Shannon et al. (2018).

A completely randomized approach may be employed (Fig. 2.15d) as in the case in some ecological sampling regimes. This involves measurements or samples being obtained at randomized locations within the study area. Achieving true random selection can be difficult due to unintentional human influence. Dividing your study area into a grid and using a random generator to select squares in which to sample is one simple approach.

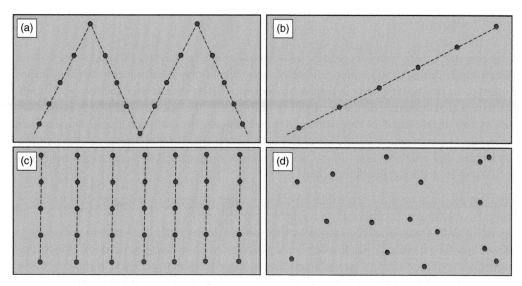

Fig. 2.15 Field sampling patterns. (a) W-pattern, typical for soil survey, (b) transect, typical for herbage composition, (c) grid, common for high resolution mapping (e.g., to support precision agriculture, and (d) completely randomized. *Source:* Sara Vero.

Two different approaches to transect sampling exist; line and point. In line sampling, a straight line of defined length is measured between two points. Everything which touches or is intercepted by the line is quantified. For example, in a species abundance survey, one would quantify what plants were at which position along the transect and also would list any stretches of bare ground (Fig. 2.16a). The point transect approach is similar, however only the species at a predetermined distance along the line is measured, for example, every 20 cm (Fig. 2.16b).

(a)

Species	Distance from start (cm)	Notes
e.g. *Bromus inermis*	5 cm	
e.g. *A. gerardii*	6–9 cm	

(b)

Distance from start (every 10 cm)	Species	Notes
0	*A. gerardii, Bromus inermis*	
10	Bare ground	
20	*A. gerardii*	
30		

Fig. 2.16 (a) Line transect record card. (b) Point transect record card. *Source:* Sara Vero.

Event Sampling

Event sampling is perhaps most common in hydrology and related fields. This approach involves taking samples during or subsequent to specific target conditions. This frequently involves measurement of surface water quality in response to storms, which generate overland flow and runoff of potential contaminants such as sediment or pesticides to surface water, although subsurface leaching and preferential flow may also be triggered under such conditions. Event sampling will involve fairly typical techniques; the key difference is in preparedness and timing. As events can be both sudden and relatively brief, if you are not completely prepared in advance you may miss them. There are some steps which can help:

- You can set email or text alerts either linked to weather stations or dataloggers at your site or linked to meteorological forecasts from reliable sources.
- Have a grab-bag of tools and equipment so that you can deploy to the field at short notice.
- Know how long it takes for you to safely get to your site, and how long you need to conduct your sampling. Your alert must be far enough in advance to allow you to get there. If it takes 36 h to reach your site, for example, a 24-h weather warning will be of no use as it will be impossible for you to reach the site in time to capture the event.
- Keep your vehicle fueled.
- If your site is distant or if you have a good general estimate of when a future event might occur (e.g., which month typically receives most storms), consider deploying to the field or to a nearby location in advance. It may be more effective to stay at a hotel near the location and limit the risk of missing the event.
- Be safe! Event sampling may pose elevated risks for several reasons that should be considered when preparing your hazard and risk assessment:
  - Weather might be cold, wet, and windy, making driving conditions poor.
  - Stream or river discharge will be high, elevating the risk of being washed away, drowning, and slips or trips.
  - You might feel rushed and take more risks, be less thorough, and feel more stressed or distracted.

Grab Versus Composite Sampling

Grab sampling is when a sample is taken at a specific location and single point in time that provides a "snapshot" of that specific moment. Frequently, grab samples are subject to environmental conditions prior to sampling and may be strongly influenced by incidents occurring recently. However, grab sampling has a valuable role in environmental research. It can be used in scoping studies to indicate the suitability of a site for future research or to evaluate conditions subject to an event. Grab samples may be incorporated into case studies or site characterization, which can provide useful additional background data to support analysis. Heterogeneity is the enemy of grab sampling! Where sites are highly heterogeneous, an unstructured approach to characterization will rarely yield an accurate understanding. To make matters worse, if there is a significant element of sorting, you may be predisposed to sampling a specific component. Therefore, the more heterogeneous or dynamic a site, material or process is, the less suitable grab sampling is as a method of characterization. You should also consider the type of heterogeneity; compositional or distributional.

Compositional heterogeneity is when a population is made up of several different components or elements, which can be in equal or differing proportions, but which is mixed. Consider a meadow in which grass and flower species are randomly distributed with no particular structure. In this instance, a single quadrat of 1 m² cut at any location in that meadow would be likely to harvest a variety of different species, however, because of randomness could not be guaranteed to reflect the overall composition of that sward.

Distributional heterogeneity is when a population of different components is structured or distributed. Imagine if the same species from the meadow were separated out and each species planted in rows of only their kind. In that instance, a quadrat cut at any location would only harvest a single species, and definitely would not reflect the overall sward composition. Distributional heterogeneity also applies to variations which occur over time, for example, diurnal fluctuations.

Both types of heterogeneity are poorly captured by grab sampling. Another example is loads in a watercourse. If the concentrations are relatively stable (i.e., do not fluctuate depending on flow or time), then grab sampling should be an acceptable indicator of water quality. However, if there is heterogeneity, such as what may arise from diurnal nutrient discharges, or if there is a dilution effect during high-flow periods, then a grab sample could not reliably indicate quality or loads. However, a modified approach to grab sampling can be used to provide a synoptic snapshot; in other words, a more thorough picture of a scenario at a specific point in time. An example of this is taking a large number of water samples across a watercourse at the same time or within a very short period (hours) during which flow is stable. Each sample remains discrete and each are non-replicated, but viewed altogether, can be used to investigate patterns across a watershed.

Composite sampling consists of multiple samples taken over a period of time or across an area. In other words, composite sampling is essentially incorporation of multiple grab samples and treating this aggregate as a single unit. It is important that representative amounts of each grab sample are present in the composite, otherwise it will be biased. A composite approach may require greater effort but has the advantage of being more representative of the area or process. Examples of composite sampling are:

- Taking multiple samples of surface soil, mixing thoroughly and analyzing for extractable phosphorus to indicate the fertilizer requirement of that field.
- Taking hourly samples of streamwater using an autosampler (Fig. 2.17) to assess total load of nitrate; due to diurnal fluctuations, large grab samples will not be reflective (see Facchi et al., 2007).

Fig. 2.17 The autosampler shown here uses a pump to extract water samples from the stream below at scheduled or flow-weighted intervals. This allows grab or composite samples to be taken without the presence of the researcher. *Source:* Sara Vero.

Replicated sampling also involves taking several discrete measurements or samples, but unlike composite sampling, each replicate is treated as an individual and is not mixed with the others. The statistical difference between the replicates should be assessed.

Sampling for Laboratory Studies

Fieldwork doesn't necessarily mean that the experiment itself is conducted outdoors. You may need to collect materials for use in controlled, laboratory tests. For example, studies of microbial respiration using incubation chambers need substrate such as soil, sediment, manure, etc. Unlike agar, purified water, or any other common laboratory material, you need to find, sample, and retrieve these samples from the field. The advantages of these studies include the ability to control microcosms to a degree which is largely impossible in situ, the opportunity to apply multiple treatments and the potential for replication and subsampling. Furthermore, you are not as logistically constrained as when you must return to the field for repeated sampling. Sounds ideal! However, you should not underestimate the challenges.

Let us take an example. You want to investigate CO_2 emissions from a range of soil types treated with a commercial amendment. Each incubation chamber requires 150 g dry weight of soil and you want to apply three rates of amendment plus an untreated control. You plan on using four replicates and four different soil types. This means you need 64 chambers, each with 150 g dry soil, or 9.6 kg. That doesn't seem like a lot but consider that you are not sampling dry soil in the field. You will collect the substrate by digging blocks of soil from 0–15 cm, removing the root mat, and aggregating. Samples will be air dried, sieved to 2 mm, and finally packed to a specified bulk density in the chambers. First, consider that field soil is not equivalent to dry soil as it contains both solid mineral particles and pore water (typically up to 25% or 30%, depending on soil type and recent weather). Therefore, much of the sample you collect in the field will actually be water, which is of no use to you. Second, you do not need the large particles, roots, rocks, and pebbles, which will be sieved out. The proportion of your sample which will be made up of this surplus depends entirely on your soil types, but suffice to say, you will need to collect far more soil than the <10 kg dry weight equivalent that will actually fill your chambers. You will need a suitable vehicle for accessing the fields, sampling tools and containers, manpower, time, and all safety equipment, PPE, and other gear which you need when going to the field. Of course, this is just one example. You might be collecting vegetation samples for cultivation in a glasshouse, trapping wild animals for tagging and release; it all depends on your unique research. However, the same basic principles of fieldwork should be adhered to throughout sampling, even though the actual experiment will be conducted back at the laboratory. In summary, you should be equally prepared for the logistical challenges even if you are not running field trials or setting up observatories.

References

Cassler, M.D. (2014). Fundamentals of experimental design: Guidelines for designing successful experiments. *Agronomy Journal* 107(2), 692–705.

Facchi, A., Gandolfini, C. and Whelan, M.J. (2007). A comparison of river water quality sampling methodologies under highly variable load conditions. *Chemosphere* 66(4), 746–56.

Glaz, B. and Yeater, M. (eds). (2017). *Applied Statistics In Agricultural, Biological, and Environmental Sciences.* Madison, WI: ASA, CSSA, and SSSA.

Howden, N.J.K., Burt, T.P., Worrall, F., Whelan, M.J. and Bieroza, M. (2010). Nitrate concentrations and fluxes in the River Thames over 140 years (1868–2008): Are increases irreversible? *Hydrological Processes* 24, 2657–2662.

Langton, S.D. (1990). Avoiding edge effects in agroforestry experiments; the use of neighbour-balanced designs and guard areas. *Agroforestry Systems* 12(2), 173–185.

Lindenmayer, D.B. and Likens, G.E. (2010). The science and application of ecological monitoring. *Biological Conservation* 143, 1317–1328.

Oehlert, G.W. (2010). *A First Course in Design and Analysis of Experiments.* Saint Paul: University of Minnesota.

Peterson, R.G. (1994). *Agricultural Field Experiments: Design and Analysis.* Boca Raton, FL: CRC Press.

Shannon, D.K., Clay, D.E. and Kitchen, N.R. (eds). (2018). *Precision Agriculture Basics.* Madison, WI: ASA, CSSA, SSSA. doi:https://doi.org/10.2134/precisionagbasics

3

Preparation

Site Selection

Where Do I Need to Look?

This may seem like a redundant question, but in fact it is anything but. If you are investigating, for example, the contribution of domestic wastewater to contamination of Lake Ontario, Canada, then at a macroscale, the answer is simple – you will need sites at Lake Ontario. However, most field-work (with the possible exception of survey) is typically conducted at a localized scale, such as at a catchment, farm, field, or plot (Fig. 3.1). The site or sites which you select must be representative of the overall area you are commenting on. Representativeness is a common question raised by reviewers. You should be able to provide a thorough and justified selection process with clear criteria for site selection. Talking to other researchers outside of your own project can be helpful in evaluating potential sites, as they may make observations or raise questions which you have not thought of. Think of this as an informal peer review and repay the favor!

Your site should be within a distance that you can travel within a reasonable timeframe and at an affordable cost, or where you can deploy to for the duration of your field campaign (Fig. 3.2). Field research has been conducted all over the world, from arctic environments to remote jungles to high elevations, so no site is impossible. However, how much time and funding do *you* have? This may restrict potential sites or limit the amount of time you can spend on location. If you can afford it and if the site is suitable for your research, don't be discouraged by remote or challenging environments. These can be opportunities for exciting research and exploration.

On a smaller scale, you should consider site access. Is there a locked gate which you need to pass through? How far and over what terrain will you have to carry equipment? Will the site be available for the intended duration of your study? These should be considered when narrowing down your site selection, for example, when choosing which field to use on a particular study farm. As an example, I once assisted a graduate student investigating contaminant concentrations in well water. The well was located across two fields of barley, and we could not drive across the farmer's crop. There was also a locked gate to which we did not have the key, and to make matters worse, the student needed huge volumes of water for each sampling round, due to the very trace quantities of contaminant present. Although the well was "ideal" on paper (located in an arable catchment, in suitable geology, near enough to the laboratory for analysis within the hold time), the reality of collecting those samples was impractical. Visiting the site prior to initiating the campaign

Fieldwork Ready: An Introductory Guide to Field Research for Agriculture, Environment, and Soil Scientists, First Edition. Sara E. Vero.
© 2021 American Society of Agronomy, Inc., Crop Science Society of America, Inc., and Soil Science Society of America, Inc. Published 2021 by John Wiley & Sons, Inc.
doi:10.2134/fieldwork.c3

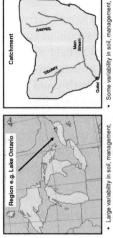

Macro-scale

Micro-Scale

Region e.g. Lake Ontario
- Large variability in soil, management, demographics etc.
- Some precipitation variability
- Suitable for large scale case study or survey but expensive and demanding

Catchment
- Some variability in soil, management, demographics etc.
- Suitable unit for land, water or wildlife management
- Suitable for case studies, surveys and monitoring
- Difficult to apply treatments or control variables

Farm
- Detailed characterisation achievable
- Suitable unit for detailed management and control
- Suitable for case studies, survey, monitoring, or installation of plot studies
- Collaboration with landowner essential
- Multiple study farms can provide representation of a production system

Plot
- Comprehensive characterisation possible
- Suitable for evaluation of treatments and controls
- Well suited to statistical analyses
- Representative of that specific site – must be justified when inferring conclusions to a wider area
- Can manipulate replicates

Pot
- Great control over variables, replication and treatments
- Opportunity for strong statistical analysis
- May be less reflective of natural conditions

Micro
- Microscopy or other high resolution instrumentation (x-ray diffraction, magnetic resonance imagery etc.) allows extremely high characterisation of individual samples
- Not deployed in field but can be used to investigate samples in the lab
- Expensive and requires specialist equipment

Fig. 3.1 Macro- to micro-scale site factors and uses.

Fig. 3.2 Consider the distance you will need to travel to reach your sites and plan accordingly. Remember, long journeys may need extra stops or additional drivers. Tiredness is a major risk for road travel. *Source:* Sara Vero

is critical to prevent these difficulties from occurring. If a preliminary visit is impossible, for example, if you are planning research abroad, then you should liaise with someone on-site or nearby.

In some cases, you might take a relatively subjective approach to site selection or specific sites may be obviously preferable for a logistical or characteristic reason. Alternatively, you might design a ranking system. These approaches assign scores based on site characteristics. This allows you to rank several sites and select objectively between them. It is particularly useful to refine down a large set of potential locations (e.g., Fealy et al., 2010). This is referred to as "multicriteria decision analysis" or MCDA.

Risk Assessment

Risk assessment is a statutory requirement under workplace health and safety legislation. As with lab work, it is essential to perform a thorough risk assessment before embarking on a field campaign. Completed risk assessments should be circulated to all staff partaking in the fieldwork.

To conduct an assessment, you need to first understand two things; hazard and risk.

A hazard is anything which has the potential to cause harm or injury to an individual, animal, or the environment. When it comes to fieldwork, hazards may be implicit to the environment you are working in (e.g., waterbodies are always hazardous because there is the potential to drown or to contract a waterborne disease, Fig. 3.3) or they may be present because or as part of your fieldwork (e.g., shears used for harvesting plant samples are a hazard because they may cut you). The first step of risk assessment is to produce a thorough list of the hazards you have identified in the area or which you or others may come in contact with (Fig. 3.4).

If you are operating in a new environment, it is a good idea to consult with someone who is familiar with it. From my own experience, working on Konza prairie during my post-doc, I was aware that I should take precautions against biting insects. However, coming from Ireland, where these are a relatively minor threat, I had not considered the very significant hazard posed by the Lone Star tick (*Amblyomma americanum*), which is numerous on the prairie and is a vector for an acquired anaphylactic meat allergy as well as bacterial infections. These were a more severe hazard than the castor bean (*Ixodes ricinus*) and hedgehog (*Ixodes hexagonus*) ticks that I was accustomed to identifying. Thankfully, the team at Konza Prairie Biological Station thoroughly briefed me on local hazards and taught me to recognize the Lone Star variety. Make the effort to seek out additional knowledge on the environment you are working in; you may not have the experience to recognize the dangers around you. You must consider the hazards specific to your scenario. Some common fieldwork hazards are listed in Table 3.1.

Fig. 3.3 Researchers in this photo lower the risks they encounter by wearing PPE (waders), working in a team and by conducting a hazard and risk assessment before beginning fieldwork. Consider what hazards you will encounter based on your site and your methods. *Source:* Krista Keels.

Fig. 3.4 Specialist training is required for uniquely hazardous tasks and environments. Managed fire, for example, requires meticulous planning and liaising with safety crews and local authorities to prevent danger to the researchers, the landscape and with civilians. Seek out training and advice from your research institute before attempting potentially hazardous activities. *Source:* Krista Keels.

Table 3.1 Some common fieldwork hazards.

Environmental	Experimental	Personal
• Darkness	• Use of vehicles	• Lone working
• Weather – cold, heat, sunlight, and rain	• Use of machinery	• Out-of-hours
• Slips, trips, and falls	• Use of tools	• Zoonoses
• Animals – livestock and wildlife	• Sharp implements	• Dealing with the public
• Water – drowning/immersion	• Electrical equipment	• Health – diabetes, allergies, etc.
• Electricity cables and buried pipes	• Chemicals – toxic, corrosive, explosive, etc.	• Traffic/vehicles
• Altitude	• Biological hazards	• Crime
• Collapsing/falling materials	• Radiation	• Exhaustion
• Loud, percussive or persistent noise		• Becoming lost/disoriented
• Fire – wildfire or prescribed		• Manual handling – heavy, awkward, or prolonged loading

The second aspect to understand is "risk". Risk is a combination of the likelihood of harm occurring as a result of a specific hazard and the potential severity of that harm (Fig. 3.5). Let us consider a scenario:

You are investigating sediment dynamics in a large river and you need to measure discharge rates during high-flow periods. This means you will be working in damp, slippery conditions, near the river, and lighting may be dim, depending on the weather and time of day. Hazards here include

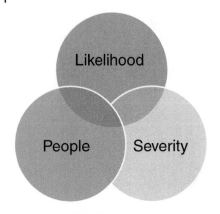

Fig. 3.5 Factors influencing the degree of risk associated with a hazard. *Source: Sara Vero.*

slips, trips, and falls on the bank or falling into the river itself. The risk of slips on the bank is moderate. Yes, it certainly could happen, but the severity of injury is likely to be low. You will not be greatly harmed, or for a particularly long time. Conversely, the likelihood of falling in the river is low – you are both careful and trained. However, should that occur during peak flow you could be washed downstream, drown, or simply become damp, cold, and exposed. Although the likelihood is low, the potential severity is great. The risk here is therefore greater than for a minor trip on the bank. You should make note of both hazards and levels of risk in your assessment.

Let us consider "likelihood" more closely. Imagine you as a river investigator are taking water samples in an area where kayakers have previously contracted leptospirosis. This is quite common in rivers but may have severe physical effects ranging from flu symptoms to kidney damage. If you have only a few samples to take and have limited exposure to the water, the risk is lower than if you have very many samples to take and a great deal of contact. The first step in reducing risk is always lowering your exposure to the hazard. What is the best way to avoid falling off a ladder? Don't climb the ladder.

The third consideration in determining risk is the individual. Certain hazards present a greater danger to certain individuals who are vulnerable. This means that they are either more likely to come to harm (e.g., someone with an existing back condition may be more likely to become injured when lifting equipment) or the potential harm is more severe (a pregnant woman exposed to a disease vector or to certain chemicals might have her health or that of her child compromised). Immunocompromised individuals or those with pre-existing conditions might be at greater risk of exhaustion or ill health during fieldwork. This should not preclude the person from taking part in fieldwork but is something that should be recognized and accommodated for the benefit of the individual and the project. It is not discriminatory to recognize and accommodate the different abilities and requirements of individuals on your team. It is responsible and helps you to involve everyone and to use their skills and capacity to best effect. Healey et al. (2001) discusses supports for disabled students in geography fieldwork. While that manual focuses on fieldwork as a teaching tool, the principles of identifying and evaluating the barriers to participation, selection of suitable field sites, thorough planning and briefing, and prioritization of health and safety are equally applicable to the research scenario. Age may also be a factor in risk. While it is certainly not always the case, older staff or faculty may be less physically capable than younger staff or students. They may be more vulnerable to accidents or may simply become tired by physical labor. Keep this in mind regarding your own safety in the future. Just because you might have thrived on week-long field excursions or could carry heavy, awkward equipment when you were 21 doesn't mean that you necessarily will a few decades later!

Different approaches are used for characterizing risk. Some use a quantitative system or scale from 1 to 5 or 1 to 10. These allow fine differentiation between levels of risk, but can be confusing for other individuals who did not conduct the assessment. Does the scale go from low to high (like burn degrees) or high to low? What are the practical implications of a 9 versus a 10 on a 10-point scale? Alternatively, a qualitative approach can be used indicating the degree of risk. While many versions are used, probably the most common is to classify risks as "Low," "Moderate," or "High." This approach is often used in workplace environments as it is almost immediately understandable to most readers of the assessment. Often a color code or traffic light system can be used in conjunction with this qualitative assessment – green for low, amber for moderate, and red for high. An example of replacing a numerical ranking system with an alternative is the visual analog scale (VAS) used in medicine. In this approach, the patient is presented with a scale of simple "faces,'" which have increasingly pained expressions. The patient simply identifies which one best expresses their level of pain. In addition to the many advantages to VAS (it is quick, requires little training, can be used by nonverbal patients), for risk assessment, a similarly non-numerical approach (such as a "traffic-light" system) facilitates communication of risk levels among team members and third parties.

Risk-assessment templates (Fig. 3.6) are available from most research institutes, and you should ask whether training is available on how to conduct them. Here is a basic, generic template which can be used for field risk assessments; however, I would strongly recommend that you prepare the Risk Management Template specific to your scenario.

Research Project:			
Project Lead:		Fieldwork Lead:	
Location/Site Address:		Co-ordinates:	
Site Contact:		Fieldwork Date(s):	
Site Description:	*Include details of landscape (e.g. rivers, forestry, mountains), distance from habitation, weather considerations, habitats etc.*		
Work Plan			
Potential Hazard	**Risk Level**	**Control Measure/PPE**	
	High ☐ Moderate ☐ Low ☐		
	High ☐ Moderate ☐ Low ☐		
	High ☐ Moderate ☐ Low ☐		
(add lines as required)			
The undersigned acknowledge the hazards and risks identified in this document and commit to, at all times, comply with and adhere to the control measures listed. All work will be conducted in line with best practice and legislative requirements. Hazards and risks may change and will be reassessed ongoingly.			
Personnel:	Name: Mobile Number: Email: Signed:		
	Name: Mobile Number: Email: Signed:		
	Name: Mobile Number: Email: Signed:		
	Name: Mobile Number: Email: Signed:		

Fig. 3.6 Generic hazard and risk assessment template. This should be edited to your specific needs. *Source*: Sara Vero.

Once you have identified the hazards and evaluated the risks, the next stage is to put in place mitigations or protective measures. Of course, it is not always possible to wholly avoid a hazard. After all, you do need to go out and take samples, investigate, and examine. This is the purpose of fieldwork. Otherwise you could simply read textbooks to learn all you need to know. So, acknowledging that some exposure to hazards is often unavoidable, what else can you do to lower the risks? Let us consider some examples:

Minimize exposure – You may not be able to avoid a hazardous activity entirely, but perhaps you can do it less. A good example here is exposure to harsh weather conditions. Perhaps monitoring equipment linked to email alerts or an app on your phone could be used to minimize the frequency that you need to venture out to the array.

Make the hazard itself safer – Imagine you need to use some heavy equipment, which will be challenging to lift and could cause injury in doing so. Could you break it down and transport it in parts so that the risk of back injury is lowered? Or could you use a trolley to roll it so that it does not have to be lifted?

Use personal protective equipment (PPE) – PPE is an essential component in safety; however, it should not be the only component in your approach to safety. In other words, it is better to remove or avoid the hazard entirely if possible, but you should also use clothing or equipment which lowers the risk of harm. I won't list all PPE here as it is specific to the task, but a few items which field researchers should consider are appropriate gloves (latex, nitrile, thermal, reinforced), waterproof clothing, footwear (waterproof, steel-toe, non-slip), hats, caps, or helmets, and lifejackets if working in or near water (Fig. 3.7). Make sure your PPE is suitable *for you*. If it doesn't fit, is uncomfortable, is damaged or worn, or for any reason is unable to fulfill its purpose, you are not protected. It is surprising how often poorly functioning or damaged PPE is used. Check your gear before you set out. Some PPE such as lifejackets or breathing apparatus require servicing at prescribed intervals. Make sure your PPE is maintained.

Use equipment correctly – It is absolutely essential that machinery and equipment are used in accordance with the manufacturer's instructions. Failure to do so may make the activity riskier

Fig. 3.7 While investigating lake sediments this researcher is wearing a life vest, wide brimmed hat and sunglasses. The latter can be easily overlooked but protect his eyes from glare reflecting off the water. What hazards might you be unaware of in your work? *Source:* Katie O'Reilly.

(increase the likelihood of an accident) or may introduce a hazard that would not otherwise exist. For example, a petrol brush cutter typically has a reinforced guard around the head. Removal or failure to secure this part can cause debris to fly or can allow the cutting wire to become dangerously long. Never bypass a safety feature or procedure for the sake of convenience or out of laziness. Test that safety features are correctly enabled before using any equipment (Fig. 3.8).

Fig. 3.8 Personal protective equipment should be chosen based on the task and equipment. The researcher shown here is wearing armoured gloves, reinforced shoes, ear protectors, safety glasses and a hard hat while being trained in chainsaw use. *Source:* Krista Keels.

Emergency

You should always have a thought-out and written plan for how to deal with emergencies, should they occur. This plan should be discussed with your team and everyone must be clear as to the appropriate procedures. You should:

- Be aware of rescue or emergency service contact details (particularly if operating in remote areas or hazardous environments).
- Establish a "buddy-system" as discussed later in this manual (Chapter 4). This helps partners to check each other's safety, raise the alert when something goes wrong or report accidents and near-misses.
- Carry identification and In Case of Emergency (I.C.E.) details such as next of kin or other contact person.
- Carry critical personal information such as allergies, blood type, donation details, etc.

You can do this by carrying and providing your team members with cards or sheets like the following template:

Personal Details

Name: _____   Research Institute: _____

Blood Type: _____   D.O.B.:____/_____/_____   Allergies: _____

Medical Conditions: _____

General Practitioner (Name & Phone): _____

Policy Provider: _____   Policy Number: _____

Phone Number:_____

Emergency Contact Details

Name: _____   Phone Number: _____

Relationship to Individual: _____

Equipment

Tools

The tools you require for fieldwork tend to be very specific to the type of sampling that you are doing. Once you become familiar with your typical field sampling protocols, you will generally know what to bring for a given day. However, in the early stages, and particularly if you are using techniques or instruments that are new to you, it is all too common to arrive at the field only to discover some essential tool is back at the office. It is a good idea to take some time to consider in detail what you might need before you set out. I do this by making a detailed list for each of my planned tasks and the items I will need for each (Table 3.2).

Table 3.2 Example tool/equipment list for fertilizer plot study sampling.

Task	Equipment	
Grass/herbage sampling	• Quadrat	• Sample bags
	• Electrical shears	• Permanent pens
	• Spare batteries	• Safety gloves (armored)
Soil sampling	• Soil corer	• Sample boxes
Pore-water sampling	• Luer-lok syringes	• Stoppers
	• Syringe caps	• Vials
Treatment application	• Fertilizer (preweighed)	• Watering cans
	• Slurry (tank to arrive)	• Weighing scales
	• Buckets (×6)	• Clean water tanks (×4)

This equipment or tool list is specific to my intended tasks and you should prepare your lists based on your objectives for the day. If your equipment has multiple parts (or essential accompanying items), you should list each component individually and check it in advance. This is particularly true if the parts are disassembled between each use. A soil corer is not of much use if you only have the augur head and not the shaft or handle! Likewise, an infiltrometer cannot be used in the field unless you have a stopwatch. If someone is helping you to prepare, make sure they know exactly what equipment you need and if it has multiple parts – it may help to number and label them in advance.

There are some general tools and equipment that are always helpful to have available:

A sturdy multitool – Look for one which has a strong, sharp blade, Phillips and flat-head screwdrivers, and a sturdy pliers. Very lightweight models are available; however, these can be somewhat fragile. My personal preference is for heavier models with strong pliers' jaws and blades which can be sharpened (repeated use will cause these to become dull). Avoid too many "junk" features. It is more important to have reliable basics. In my experience, it is worthwhile researching a multitool that you like, and which is well reviewed. There are several major brands that are very reliable and offer lifetime warranties, repairs, and servicing. There are also models specifically designed for certain types of work (e.g., with features for mechanics or for wiring). I consider my multitool to be absolutely essential and never go on fieldwork without one.

Cable/zip ties – Next to the multitool, these are the most useful piece of equipment I can think of. I bulk buy these in different sizes and always keep a handful in my vehicle, backpack, toolbox – everywhere! These can be used to secure equipment, fasten gates, seal containers, repair things, as markers and flags, the list goes on. I have even secured a logger-box, solar panel, and antenna using only these. They are inexpensive, replaceable, and strong. If you cut a cable tie in the field, *always* collect the pieces and bring with you. Fieldwork is no excuse for littering.

Carabiner – These clips are favored by climbers and hikers and can be used to fasten tools, packs, etc. You should always have at least one – for your keys! Losing your vehicle keys in the field can be a genuine disaster. I know one scientist who accidentally buried his in a soil pit! Fasten your keys to your belt.

Sample bags (zip or press-seal) – Always have enough sample bags for your planned samples and a large number of spares. You should never need to omit a sample or miss an opportunity for an additional sample because of a lack of bags. Furthermore, depending on the weight and shape of your samples you may need to double bag them to prevent bursting. Wet soil or river sediment samples can leak water during transport if bags are not perfectly sealed, so always double-bag these. Some sealable bags feature opaque panels for writing on. These limit smearing and make it easier to read labels. If possible, label your sample bags prior to going to the field. You are likely to write more legibly in comfortable, dry office conditions, and it will also save you time.

Sample boxes and/or bottles – Similar rules apply here as for sample bags. Label in advance and always bring spares. If using sample bottles, always label the bottles themselves, not the lids.

Waterproof plastic crates – For transporting all samples, storing equipment, etc.

Duct-tape – Useful in many situations; for repairing equipment, wrapping wires, patching clothes, sealing sample bags and boxes, etc.

Permanent markers or pens – Several colors.

Measuring tape – Your job as a researcher is to be precise. Never estimate size or scale; be prepared to make an accurate measurement. Check that your tape has metric markings, not imperial. Measurements taken for research should be recorded in mm, cm, and m, not in feet and inches.

Waste bags – Like other outdoor activities, fieldwork should be conducted with the principle of "Leave no trace." Of course, you may be installing permanent or semi-permanent equipment, but aside from planned changes that are an intended part of your work, you should never alter your sites by leaving waste or debris.

Lab gloves – Various sizes and plenty of spares.

Work gloves – Sturdy work gloves can prevent blisters and cuts and help you grip slippery or rough objects. Different weights and varieties are available in hardware stores. It is also possible to get sturdy, fingerless gloves from outdoor and sporting goods stores. These can be useful if your tasks also require some dexterity.

Flashlight or headlamp – This will allow you to work in dim conditions and adds to your safety. I prefer a headlamp as it leaves hands free for working (Fig. 3.9). Remember to bring spare batteries.

Fig. 3.9 A head flashlight is a valuable piece of equipment as it leaves your hands free for taking measurements and using equipment. This researcher is setting up an accoustic device for monitoring wildlife. *Source:* Nikki Roach.

Keep these tools and/or equipment in a toolbox or bag, so that they don't need to be gathered every time you go to the field. You may add permanent items depending on what you routinely use. Having a "go-bag" is particularly useful if you engage in event sampling and need to rapidly deploy (Fig. 3.10). If you have a field-vehicle for your exclusive use, you can keep these in it at all times. Remember to replenish your toolkit regularly and also clean out old, empty, or damaged equipment. If you break, lose, or run out of any basic equipment in the field, you may be able to restock at a hardware store if there are any in the vicinity.

Fig. 3.10 Although specialist equipment varies, it is helpful to keep some basic essentials ready to go. My "grab-bag" always includes a multi-tool, rugged smartphone, zip-ties, sample bags and permanent markers. Keep your essentials in a bag or tool-box so they are always ready. *Source:* Sara Vero

Consumables

Consumables include all the single-use equipment which you will use during fieldwork, such as syringes, sterile equipment, filters, swabs, etc. Consumables may also include chemicals such as reagents and standards. Always carry spares or surplus of all consumables. There is a phrase attributed to the U.S. Marine Corp: "Two is one and one is none." This is certainly the case with consumables. You must budget for spares in the planning stages and be aware of lead times when scheduling fieldwork. If your consumables do not arrive before your intended start date, you may be forced to postpone. Bear in mind that standards and reagents frequently have use-by dates or may require chilled storage.

Maps, Apps, and Fieldsheets

There are many different types of maps which you might use depending on your field of research from topographic, road, and river maps, which any driver or hiker might have in their car or rucksack, to more specialist maps detailing geology or soil type (Fig. 3.11). If you are using these latter types, you probably will have some training in their design, use, and interpretation;

Fig. 3.11 Maps are essential for visualizing and understanding the fieldwork site. *Source:* Jaclyn Fiola

so, we will mostly address maps which you would typically use to orient yourself in the field. Maps can be used for selection of potential field sites prior to fieldwork, for planning your travel routes, for orienting yourself in the field, for marking sampling sites or other observations, etc. They are an essential tool in your fieldwork kit. To use them effectively, we must understand their limitations.

All maps, no matter how accurate or at which scale, are *simplified interpretations* or visualizations of reality. In other words, they are a symbolic representation of what we *think* we see. Consequently, misconceptions of what we observe, when recorded as a map, become part of our understanding of that environment and so; false understandings can be passed on to others. However, if we appreciate the limitations of maps in this regard, we can use them effectively.

Scale is the relationship between the distance measured on a map (or photograph) and the distance which it represents in reality. In other words, what each mm, cm, or inch on the map indicates in terms of those same units in the field. The scale itself is unitless; rather it expresses the ratio. Let us imagine you are using a map with a scale of 1:10,000. In that case, each 1 cm you measure on the map would equate to 10,000 cm (or 100 m) in the field. If you chose to measure in inches, each inch on the map would equate to 10,000 inches (or 278 yards).

You should always consider the appropriate scale as regards your intended use. Large-scale maps (e.g., 1:24,000) depict relatively small areas with a large amount of detail (e.g., USGS topographic maps). Conversely, small-scale maps (e.g., 1:100,000) reflect large areas but with lesser detail. For example, a large-scale map might be used to understand the terrain of your field site, to plot your driving route or to outline a watershed, while a small-scale map might be used to illustrate where your site is located within your state or country. If a map is photocopied and either shrunk or enlarged, the ratio may no longer be accurate. However, the bar scale (which is a visual representation of distance on the map to distance in reality) should still be reliable as it will be distorted to the same extent as the rest of the map. In other words, if the bar represents 1 km, then whatever length the bar is will still represent that same distance.

Another thing to remember is that maps are fixed in time. If the landscape changes due to natural occurrences (such as a landslide or wildfire) or human activities (such as construction of a new road or excavation of a quarry), the reality will not match what you read on your map. So be aware of the age of your map when you use it for reference material. Taking these potential limitations into account allows us to use maps appropriately.

You must know which direction your map is oriented. Typically, north will be indicated by an arrow or compass symbol. Of course, for this to be useful, you must know where north is in relation to your position in the field. A magnetic compass will tell you this.

Unless it is a sketch map (detailed below), maps will indicate the scale at which they were prepared. This allows you to calculate distance or to put into context what you are observing in the field. Scale is usually indicated by a scale bar.

Grid references are used to define coordinates on a map. There are several grid reference systems. The Universal Transverse Mercator (UTM) is used internationally, while nations such as Ireland and the United Kingdom have national Ordnance Survey (OS) systems. OS use grids of 100km^2, which are further divided into 10km^2 units. In the United States, the Land Office Grid system is used, which divides land into townships of 36 square miles and further, into sections of 1 square mile. The basic premise of grid referencing is that the mapped area is overlain with a square grid which is numbered sequentially from left/bottom to right/top (Fig. 3.12). The horizontal axis defines the "Eastings," while the vertical access defines the "Northings." The scale of the grid will vary although many maps use a 1km^2 spacing. When using a 1km^2 resolution system such as the UK Ordnance Survey, a four-figure grid reference will indicate a specific grid square, while a six-figure reference will indicate a specific 100m^2 area. Grid references are read as follows:

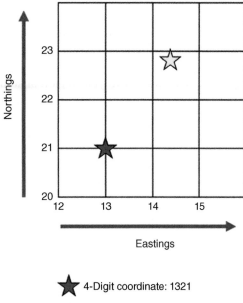

Fig. 3.12 How to find coordinates on a map grid. *Source:* Sara Vero.

1) Orient the map vertically (so that North is topmost).
2) Read the Easting first. Count across the numbers along the bottom of the grid. This will give you the first two digits of your six-digit coordinate.
3) To find the third digit, you divide that grid square into 10 equal parts and estimate the position of your object or location.
4) Now find the Northing. Count the vertical numbers along the side of the map. This gives you the fourth and fifth digit.
5) To estimate the final digit, repeat step three in the vertical direction.

A simple grid is shown in Fig. 3.12. A four-digit reference is shown as the red star, and a six-digit reference is shown as the yellow star. Six-digit references allow greater accuracy than four-digit references, so if you need to locate a precise position, they are more helpful.

Local or Regional Maps

There are several reasons you might use a map of a region or a country. First, for context. The geographical location of your sites is crucial for contextualizing your work with respect to climate, landscape, sociology, etc. When presenting your research to wider audiences who might not be familiar with your region of study (as is typically the case at major conferences or in peer reviewed publications), this type of map is a quick way to provide a lot of contextual information. The second reason is in research involving surveying. These will typically require maps at a relatively local scale, such as 1:20,000. These may be used to guide researchers to target sites or may be marked with sites or observations made by the researcher in the field. Some research involves large areas, such as watersheds spanning hundreds of square kilometers. In these cases, you may need both regional maps to show the entire area and also large-scale maps to show more detail and aid site characterization (e.g., topography, marking observations, etc.).

Sketch Maps

A sketch map is a simplified, hand-drawn map incorporating the key features of an area. They are very useful for recording the layout of a field site and can be used to guide development of more accurate maps using GIS when you return to the office. It is often difficult to produce sketch maps to a particular scale, so they should not be used prescriptively. However, if you take a measurement (e.g., the length of a roadway or ditch, or the dimensions of a building), you can note these figures on your map. While there is no fixed rule, I recommend scanning your sketch maps when you return to the office and storing as a PDF.

Steps for preparing a sketch map are as follows:

1) Write the site name, GPS location or grid reference, date of survey, and your name. This will help you or other team members to correctly interpret and use your map. It is common in the United States to use month/day/year format, while in Europe day/month/year is standard. To avoid any confusion, it may be best to record the date as text, for example 8th March 2019.
2) Consider the area you are mapping from a birds-eye perspective.
3) Outline the extent of the area (e.g., borders of field) and include dimensions if known. You can note estimated dimensions if actual measurements are not possible.
4) Find north and note on map.
5) Sketch key features such as rocky outcrops, watercourses, roads, buildings, vegetation, etc. If you include a watercourse, note which direction it is flowing.
6) Indicate slope, either with elevation measurements if possible or simply high-to-low.
7) If you can measure any dimensions, note these on the map.
8) Note where you take samples. If you take various types of samples (e.g., water, sediment, and soil samples), use different symbols or colors.
9) You can also sketch "invisible" features. For example, if you are told by the landowner that a subsurface drain runs east to west for 500 m across the southern edge of the field, you can note this on the sketch map using a dashed line, even though you did not in fact see that feature.

Here is an example of a sketch map from a recent farm visit (Fig. 3.13):

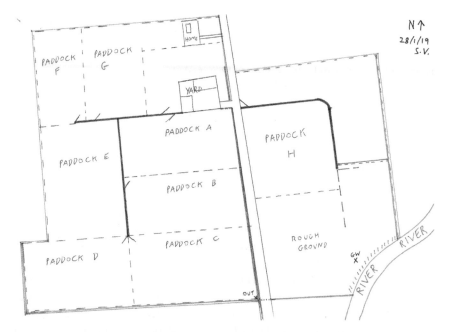

Fig. 3.13 Sketch of a farm field site (note European date format). *Source:* Sara Vero.

Plot Diagrams

Unlike typical maps, plot diagrams are not always drawn to scale, but rather, are indicative of where each treatment is to be applied (Fig. 3.14). Plots are frequently laid out in an orderly grid, so color-coded or numbered diagrams are particularly useful. These diagrams can be simply produced in Microsoft Powerpoint, Paint, or other diagram or image software.

All you need to do to produce such a diagram is to draw the appropriate number of boxes in according to the intended outline of your field plots. Using the text function, you can label each

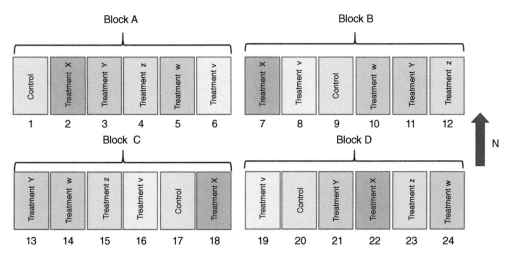

Fig. 3.14 Example plot diagram for a field trial. *Source:* Sara Vero.

box according to treatment, replicate number and any other identifier. It is often very helpful to color-code the different treatments. Finally, be certain to indicate direction on your map, as there may be little visual indicator of differences between blocks or plots, depending on the layout and stage of your experiment.

If you do not need to mark your plot diagrams and are simply using them for reference, laminate them. This prevents damage and allows them to be wiped clean.

Mobile Apps

Phone apps are becoming increasingly popular as alternatives to printed maps. These GIS-based software use GPS to aid in location. The improved accuracy and positioning allow you to orient yourself in the field and can allow "tagging" of locations so that your sample points or observations are precisely recorded. However, there are some considerations which you should bear in mind:

- GPS is notoriously demanding on battery. If you are using GPS-enabled software on a smartphone or other device, carry a portable power-bank and keep a charger in your vehicle.
- Be conscious of spatial accuracy. Most commercial handheld GPS is accurate to within 3–10 m. Keep in mind that any site you are looking for or record is only accurate to within this distance. The GPS apps on smartphones vary in accuracy depending on whether Wi-Fi or network positioning are integrated, and also depending on make and model of phone. Typically, 5 m accuracy is common for smartphones, but wide ranges are observed. High-accuracy surveying GPS devices are available, which are reliable 1 m. Note that errors in positioning rarely arise due to the transmitted signal, but rather due to blocked line of sight, errors in the map software, or indoor/obstructed environments.
- Beware of wet conditions! Apps are useless if weather conditions are too poor to expose your phone or device. If this is a possibility, you might consider investing in a waterproof or ruggedized case. If you are working in or around waterbodies, remember to check to what depth or degree of submersion the case is reliable and also, that it may not be possible to safely retrieve a device dropped into water.
- Bring a print map. If your device runs out of battery, loses signal, runs out of data, becomes damaged, wet or lost, you need to be able to finish your work. Print maps should always be available as a backup.

Field Sheets

A field sheet is used to record your notes, measurements, and observations (Figs. 3.15, 3.16, 3.17, and 3.18). For general observations, a notebook will often suffice, however, if you intend to record several parameters or sites, it is useful to prepare and print sheets (or save them to your tablet or

Fig. 3.15 This salamander is taking an interest in one wildlife researher's field sheets! *Source:* Nikki Roach.

Fig. 3.16 This researcher is taking careful notes using a pre-formatted field sheet. This helps maintain legible notes and stops crucial details being forgotten. *Source:* Krista Keels.

Fig. 3.17 A well-maintained notebook is a good alternative to printed field sheets. However you choose to record data in the field, it is best practice to back it up on computer once you return to the lab or office. *Source:* Derek Gibson.

Fig. 3.18 The buddy system in action: one reseacher is taking measurements while another is recording the observations. *Source:* Katie O'Reilly.

smartphone) with preset fields for you to fill in. There are often standard sheets that are used in particular fields of research, for example, pit cards for preparing a soil profile (Fig. 3.19). As well as recording your notes in a sensible and clear fashion for transcribing later, field sheets are a useful

Profile Description								Date:	
Surveyor:		Site ID:		Lat:	N	Long:	W	Photo #	
Slope position:				Elevation:		Slope:	%	Aspect:	
Drainage class:					Land Use:				
Landscape:					Geology:				
General Notes:									
Horizon	Depth (cm)	Structure		Texture	Bulk Density (g cm³)	Stoniness	Colour		Notes
									e.g. Features, artefacts, water, etc.

Fig. 3.19 Simple soil profile description template. Note that national soil survey and pedologists use more detailed and complex systems. This is a simple approach which should be sufficient to provide contextual information, usually accompanied by a sketch or photo. *Source:* Sara Vero.

prompt to prevent you from forgetting any measurement. If designing your own sheets ensure that you leave fields for all of the relevant information including who is taking the sample, the date, location, weather conditions, measurement values, etc. Standardized manuals for field description are also available, particularly for soil survey. These provide comprehensive pit cards and will help ensure that your descriptions correspond to industry guidelines and so can be interpreted by individuals who were not present at the time of recording. The Field Book for Describing and Sampling Soils (Version 3.0) (Schoeneberger et al., 2012) is available for free download from the National Resources Conservation Centre website. Another excellent resource is Soil Science: Step-by-Step Field Analysis (Logsdon et al., 2008).

While this might seem obvious, it bears mentioning; write clearly (Fig. 3.18)! Time is often is often wasted trying to decipher poorly written field notes. If there are multiple members of your team, agree in advance on any shorthand, acronyms or jargon you intend to use. Notes are only useful if the information can be understood afterward. When designing custom field sheets, leave extra space for additional notes or observations. It is easy to neglect noting things down otherwise, and cramped handwriting squeezed in the margins is often impossible to interpret later on (Fig. 3.15).

Photographs

Photographs are a useful method to precisely and permanently document what you observe in the field so that you can later prompt your memory, provide visual aids for presentations and in some instances, analyze for data. As high-quality digital cameras are increasingly inexpensive and available and as more people carry smartphones with impressive capabilities, it is increasingly common to incorporate photography in fieldwork. Photography should be used as an aid to memory; not a replacement. Studies have shown that reliance on photography during concerts and events has reduced the recollection of the photographers of these occasions.

There are several theories as to the cause of this; including "displacement theory" by which the photographer fails to commit to memory because they subconsciously have lodged it elsewhere; as a photograph. However, there is no information available yet as to whether this occurs in a field or research scenario. Simply, I would recommend that you maintain a high level of awareness and observation in the field and use photos for their correct purpose; as tools and aids.

A recurring issue with fieldwork photos is remembering where it was taken. Of course, you should have a logical file system on your computer back at the office with folders for individual sites, but when initially downloading your photos you will need to know where, when, and what you photographed. You can write the site location, date, and any other key information on a sheet. Photograph this and then take the photograph(s) of your site, experiment, or observation. A soil scientist recommended to me that when photographing multiple soil pits on a single farm he numbers his excavations by holding up the appropriate number of fingers. Low-tech but perfectly effective!

Download and file your photographs as soon as possible on returning from the field. Photos can be accidentally deleted or simply misfiled, and the details pertaining to that image will be forgotten. Consider this the equivalent of promptly filtering your water samples or labeling your soil cores.

Use a consistent labeling procedure for photos. Site name or ID, date and topic of the photo are helpful to include.

Geotagging is the addition of spatial data such as latitude and longitude to a digital image. This allows precise identification of the location of the photograph and enables these images to be incorporated into digital, searchable maps. Initially, geotagging was conducted by recording locations using handheld GPS devices and linking these coordinates to the photograph using desktop software. However, now many cameras and smartphones incorporate a geotagging function which correlates these data automatically. Welsh et al. (2012) observed that the relatively recent ubiquity of smartphones enables geotagging to be widely incorporated as an aid to geography fieldwork, particularly as an aid to post-field reflection. That paper also noted some limitations of the approach (chiefly cost and signal coverage); however, these may be overcome by more recent advances in mobile phone technology.

Unlike most photography, the purpose of your research photos is not to be aesthetically pleasing, but to accurately document your observation. The following steps will help you to do so:

- A tripod will stabilize your camera and allow a level photo to be taken.
- Center your photo on the feature of interest.
- If possible, ensure the feature is well lit, clear, and free of shadows. This may not be entirely within your control depending on the time of day, weather, etc. Using a lens hood may reduce glare from the sun.
- Include a tape, ruler, or object of known proportions for scale (Fig. 3.20). When photographing a soil pit, it is standard to include a tape measure with clear markers so that the depths of each horizon can be identified. If you don't have a tape or ruler available, you can place a pen, trowel, or other small common item next to the feature of your photo.
- If using a smartphone camera, hold it in landscape position.

Accurately representing the color of a soil profile or sample can be challenging. Cameras do not capture the absolute color, but rather a relative value for each individual picture. When digital photographs are transferred from one device to another (such as from the camera to a laptop) this may become further distorted. You can correct the color of digital photos using color matching software such as Color Matching Module (Windows) or Colorsync (Macintosh) (Ortiz et al., 2014).

To do this you must include a chip or card with one or more known colors (such as a primary color) in your photo. You can buy color checker cards from photography suppliers or simply use your Munsell chart (Fig. 3.21). The software will allow you to adjust the color of the soil profile relative to the reference color. If the photo is only to be used for illustrative purposes, this may not be necessary. However, if you are extracting data based on the exact color then this adjustment is essential.

When taking photos, particularly if they include people, you should make certain that it is permissible to do so. On research facilities or public land, it is generally permitted however, on private property this may not be the case. Photography might also be prohibited in and around certain buildings including airports or public offices. You should ascertain whether your team are comfortable with being photographed during their work and respect their wishes. Be respectful of the public also and particularly, never take photos of minors (even when they are not the focus of the photograph) without the express permission of their parent or caregiver. My personal recommendation would be never to do so, unless your research involves a social science aspect.

Fig. 3.20 Using a tape for scale. *Source:* Giulia Bondi.

Fig. 3.21 Munsell charts or other color checkers can be used to identify and if necessary, match the color of soils or other features observed in the field. *Source:* Jaclyn Fiola.

Local etiquette should always be considered, as some cultures consider photography to be highly inappropriate or invasive.

Social media is an effective tool for promoting your research, and it is well established that tweets and posts that involve images receive more likes, retweets, and generate more interaction by engaging the audience. Photos used for this purpose might be the "utilitarian" type used for analysis or review, but equally, can be "action" shots, showing you and your team at work or beautiful, funny or interesting things you see during the course of your work. These might have limited use from an academic perspective but can be very effective in communicating to a wide and nonresearch focused audience. The "story" feature of Instagram allows photos and short videos to be uploaded live and many researchers are using this to document "a day in the life." You might be interested in taking part in something like this, but don't allow social media to disrupt your research!

Be careful of taking or displaying photographs which might be negatively or incorrectly interpreted, or which might bring controversy or disrepute to your work. For example, ecology research might involve the capture and release, tagging, or anesthetization of wild animals. Although you will conduct your research ethically and with best practice, photos of these actions might be interpreted as "cruel" or even illegal. There is a particular risk where such photographs are sent on social media including Twitter or Facebook as the level of explanation, context, and understanding of the viewer might be low. There is no explicit rule regarding this, but you should exercise discretion if there is a potential for controversy.

If displaying a photograph which you did not personally take as part of a presentation, paper, or report, you must obtain the photographers permission and cite them appropriately. Those images are their intellectual property.

Personal Protective Equipment (PPE)

Before you embark on fieldwork, it is your responsibility to ensure that you have appropriate and functioning PPE. These can not only include clothing such as steel capped or waterproof boots, safety helmets, and gloves but may also include more specialized equipment such as harnesses, lifejackets, face shields, full body suits, etc. The risk assessment should guide you as to your PPE requirements. Assuming you have correctly identified the hazards and selected the appropriate PPE, you should also consider the following:

Is my PPE rated for the hazard in question? For example, ear protectors are designed to shield against noise up to a certain decibel, above which protection is limited (Fig. 3.22). Check the rating of your equipment and upgrade if necessary.

Has my PPE been serviced? Lifejackets and other some other PPE require servicing (typically on an annual basis) to ensure that all moving or inflatable parts, straps, and buckles, etc. are functioning optimally. If your PPE has been in storage, check its documentation for the last recorded servicing and renew as necessary.

Fig. 3.22 Check if your PPE is appropriately rated for the hazard in question. Ear protectors, for example, should be selected based on the decibels you will be exposed to. *Source:* Jesse Nippert.

Does your PPE fit correctly? Equipment which is too large or small may not offer the correct degree of protection and may be hazardous itself. Always try on your PPE equipment before going to the field.

Visually inspect your equipment. Damaged or worn equipment cannot offer the degree of protection as it is intended to. Repair or replace. Don't take risks.

High-visibility (hi-vis) clothing is recommended when you are working in potentially hazardous environments (e.g., watercourses) and around machinery or traffic (Figs. 3.23 and 3.24). It is often a good idea to wear hi-vis if you are in semi-urban areas (e.g., farmlands). This is both for your own safety and to put landowners at ease. Someone in nondescript clothing may appear suspicious or even frightening; however, your hi-vis gear identifies you as being there for legitimate purposes. Always identify yourself clearly, and unless there is prior agreement for free-access, contact landowners if you wish to enter their property.

Fig. 3.23 Hi-vis clothing can help drivers or machinery operators to see you, will help you be located in remote environments, and may help reassure people that you are there for legitimate reasons. *Source:* Sara Vero.

Fig. 3.24 Keep your PPE clean and in good condition. Store appropriately after use. Remember; damaged or poorly functioning PPE may not protect you and may create a false sense of security. *Source:* Karen Vaughan.

Vehicles

Your research institution may have vehicles available for fieldwork, from trucks and vans to all-terrain vehicles (ATVs), boats, and snowmobiles (Fig. 3.25) There are several things to consider when using field vehicles.

Do you have the skills, experience, license required to *safely* and *legally* operate that vehicle? Is training available? If you are an inexperienced or untrained driver, you should not operate a vehicle in the field. Consider the risks posed to both you and others. You must abide by the law and never operate any machinery for which you are not licensed. Don't assume that things will be ok or that a risk is acceptable. Never give in to pressure from peers or colleagues in this regard. Many field-vehicles such as quads, ATVs, or boats require training or even specialist licenses. In some counties, specific licenses are also required to tow a trailer on a public road. In such cases, check whether your institution can either directly provide or facilitate training. The importance of training and familiarity with your vehicle or indeed, other machinery, cannot be over-emphasized as injuries and sadly, fatalities do occur. Simply being able to turn on and steer an ATV, boat, or other vehicle is not the same as

Fig. 3.25 If you are using boats, all terrain vehicles, snowmobiles or other vehicles make sure that you have appropriate training and become familiar with their use prior to fieldwork. Remember that PPE may also be required. *Source:* Julie Campbell.

being competent – there is much more involved in their skilled use. It is not uncommon for individuals with a passing familiarity with off-road vehicles to have a false or overestimated understanding of both their own and their vehicles' ability to cross rugged terrain, which can lead to accidents. Experience and proper training will reduce this risk. Remember that training does not only help your present research but also helps you to become a more capable and useful individual.

Similarly, you must only ever use vehicles and machinery as they are intended. Not only does misuse of vehicles increase the likelihood of an accident occurring but it also may invalidate your insurance.

While the availability of public transport has in many places lowered the need for individuals to learn to drive, this is for the field researcher an extremely important skill. If you are working alone, how else will you get to your site? And if working in a team, can you take your share of the burden as pertains driving? What happens if the "driver" on your team is unavailable or simply too tired to drive any further? If you intend to have a career involving field research, driving is a primary skill that you should learn.

What is the insurance policy regarding the vehicle? You may need to submit a copy of your drivers' license to your institution before using a shared vehicle. If using a private vehicle, you should arrange appropriate insurance which covers use for work purposes. If you are using a rental vehicle and intend to take it off-road, check whether the insurance policy covers such use as this may in fact void your cover.

Is the vehicle suitable for the terrain? If you are simply driving on public roads and highways to a site most cars, vans, or pickups are capable. However, if you are driving on small or temporary tracks or trails, traversing off-road or across fields, bush, or gravel, you may need a more rugged vehicle (Fig. 3.26). This can mean an entirely different vehicle (such as a four-wheel drive) or simply different tires for improved traction. Pickups are often desirable for fieldwork; however, these are often "light" in the rear which can allow fishtailing or skidding on loose or uneven surfaces. Drive slowly, in a low gear in these circumstances and consider weighting the rear of the vehicle using equipment or sandbags.

Is the vehicle suitable for the weather? Even if your truck is normally capable in a certain environment, under inclement weather conditions, when the ground is wet or icy or if there is flooding it may be unsuitable. ATVs are often more capable, although they offer minimal shelter for the driver and may necessitate multiple trips if transporting machinery. Whatever vehicle you use, always select the appropriate tires for the season and weather (Figs. 3.27 and 3.28).

Is the vehicle suitable for carrying your equipment, samples, and team? It is remarkably easy to fill even large vehicles with equipment and samples very quickly. Before going to the field load the vehicle fully so you can see how much space will be available. Depending on the type of samples and equipment, you may need a vehicle which has a separate cab, or a divide between the human and storage areas. Examples of this are samples stored in dry ice, samples which pose a biohazard, or samples which are simply too smelly to reasonably transport in close proximity to the driver (e.g., manure or slurry samples)!

Is the vehicle shared by other people? Institutional vehicles are frequently in high demand, particularly during seasonal fieldwork. Usually, either a physical or digital calendar is maintained for shared vehicles, which will allow you to book in advance. This is essential; imagine postponing your carefully planned fieldwork because you forgot to book a vehicle! Be conscientious in your bookings and encourage responsible behavior in your colleagues. By this I mean, if you cancel or postpone fieldwork for any reason, correct the calendar. Perhaps someone else can

Fig. 3.26 All-terrain-vehicles can be helpful for accessing difficult to reach sites or carrying bulky or heavy equipment. Some researchers develop novel alternatives, like this impressive off-road bike and cart. Here, it is being used to carry auto-sampler carousels and water sampling equipment. *Source:* Lauren McPhillips.

Fig. 3.27 Vehicles can get stuck in snow....... *Source:* Andrea Brookfield.

make use of the vehicle on that day. Likewise, return the keys on time and leave the vehicle fueled, clean, and in good condition. If you cause or become aware of any mechanical issues, make the person responsible for vehicles aware at the earliest opportunity. Selfish or thoughtless behavior regarding shared vehicles can make a person very unpopular with their peers, and even result in banning from use.

Fig. 3.28 or soft ground/mud. *Source:* Brandon Forsythe.

Where do you fuel the vehicle? Your institution will likely have either facility for refueling or a policy/procedure for fueling at public locations. In the latter case, you should keep the receipt and submit it promptly via your expenses system. Alternatively (or if using your private vehicle), there may be a flat rate per mile/kilometer at which you will be reimbursed. Use your odometer to keep track of your accumulated mileage.

A brief comment on using institutional vehicles; in many cases, they will be identifiable by colors, decals, or logos. When using these vehicles, you are representing your university or employer, and misdemeanors, irresponsible, or dangerous behavior will reflect poorly and rapidly be reported. This can be embarrassing for you and for your institution and can lead to disciplinary action. You should always abide by the rules of the road, drive and operate safely and responsibly in the field and act courteously in public areas such as rest stops, service stations, and garages.

Load your vehicle carefully (Fig. 3.29). This will make fieldwork much easier. If possible, arrange equipment so that you can easily access those items which you will need frequently or first. Keep storage containers upright, using crates or boxes when necessary. You can use straps to tie down and secure loose equipment; this is particularly important if you have an open pickup or if there is no partition between the storage area and the passenger cab. Given the option, it is usually safer to have separate cabins for equipment and personnel.

A final important consideration; drowsiness has similarly detrimental effects on brain function and road safety as alcohol (Vakulin et al., 2011). Fieldwork can involve long and exhausting days. You will likely be more tired than a day spent in the office or laboratory. The European Union rules that commercial drivers do not exceed 9 h per day, or 56 h in a week and that a break of at least 45 min be taken after no more than four and a half hours. Stop the vehicle in a safe place such as a rest area and take a nap, eat something, and drink something. Do not drive when you are overly tired. Either swap with a member of your team or stop overnight if necessary.

Fig. 3.29 This van has been expertly packed for a field trip. This will prevent damage during transit and makes it easier to deploy your equipment efficiently when you arrive on-site. *Source:* Karen Vaugan.

Weather

Target Conditions

Target conditions refer to the weather that you either want or need to conduct your fieldwork. Let us consider these as either **preferred** or **essential** conditions.

Preferred conditions are those which are ideal for ease and efficiency of fieldwork. They are generally broader than essential conditions; in other words, if you are maintaining a grass plot study, you may prefer to apply a fertilizer treatment on a relatively dry, cool day. However, if it is a scheduled treatment, there may be moderate flexibility (a couple of days) around the timing of application, but ultimately you cannot wait for particular conditions and must adhere to your plan even if that means working during rainy or cold weather.

Essential conditions are those which must occur for your fieldwork to be conducted. If you are studying soil erosion during rainfall, there is little point in deploying to the field on a hot, dry summer day! You must define in advance what conditions are required – how much rain, what air temperature, what river flow, etc. Consider both your own hypotheses and the literature to determine these targets. Depending on how reliant you are on specific conditions, you may need to have a contingency plan for if conditions are not suitable or if they change suddenly. For example, if you are "event sampling" from a river, you need relatively high rainfall to generate overland flow and increased discharge. Should rainfall be insufficient, you may need to postpone your sampling, even if the initial forecast was more favorable. Consider the reliability of forecasts in relation to this. The strength of predictions typically increases as time is shorter; in other words, a 24-hour forecast is more reliable than a 48-hour forecast. You can make provisional plans based on longer term predictions but be prepared to adapt based on the short-term forecast. Since event sampling is often subject to narrow windows of opportunity, it is sensible to have your equipment ready in advance so that you can deploy to the field at short notice. This can be challenging if your site is remote. If conditions change during fieldwork, you may need to terminate the operation. This can be a difficult decision, especially when time, effort, and expense have already been committed. However, as a researcher your primary goal is to collect the data to allow you to test a hypothesis. If the weather conditions compromise this data, you should not continue. Good data can be collected another time. Bad data are often a greater waste of resources.

Precipitation

Rainfall, sleet, hail, and snow obviously have a significant influence on both the environment and depending on the severity, your ability to conduct fieldwork successfully. For some studies, such as examination of runoff or streamflow, precipitation above a certain magnitude can be characterized as an "event," and you may wish to target your measurements within this period. In such cases, preparation in advance is crucial, as events may be transient, and you can risk missing them entirely if you are not attentive to the forecast. It is particularly risky where you have long travel times to your site. The accuracy of rainfall forecasts varies by climatic region, forecasting agency and time (i.e., the more distant the estimate is made from a given date, the greater the likelihood of error). Interestingly, surveys conducted by Joselyn and Savelli (2010) revealed that the public believed rainfall to be commonly underpredicted by meteorological agencies, despite high accuracy and low bias in reality. The best approach for a researcher is (i) to trust meteorological forecasts – they are made by experts, (ii) to consult several forecast agencies, (iii) to set up phone or email weather alerts, and (iv) to be prepared in advance to deploy to the field when an event is incoming.

At the other end of the spectrum, rain or snow can be an impediment to your fieldwork as it dampens field sheets, worsens ground conditions, and makes driving more hazardous. There are no strict rules as to whether you should go ahead with fieldwork under damp conditions; it depends on what you are doing and the severity of the weather. These questions should help you to decide:

1) Can I safely travel to and access the site?
2) Can I still take reliable measurements, or will they be impaired?
3) Am I likely to damage or influence the site (e.g., in plot experiments)?
4) Can I take additional time as I might be slower?
5) Can I postpone until there are better conditions or is timing critical?

If you go ahead with fieldwork, a few things can help you to deal with inclement weather:

- Wear comfortable, well-fitting, waterproof clothing (raincoats, waterproof over-trousers, waders, etc.).
- Bring towels and spare dry clothing to change into afterward.
- Wear suitable footwear – high, waterproof boots.
- Always bring spare socks. (This is crucial in my opinion!)
- Allow extra time in the field and accept that you may not get as much done as usual.
- Drive safely and allow extra travel time.
- Bring a weatherproof clipboard to shield field-notes or use a ruggedized laptop if available.
- Bring waterproof boxes or crates to store equipment and samples.

Humidity and Temperature

Humidity (relative) is the ratio of the partial pressure of water in the air relative to equilibrium and is typically expressed as a percentage (Fig. 3.30). Humidity can be uniquely challenging to outdoor work, both at the very high and very low ends of the scale as it impacts our ability to moderate body temperature. This has been documented in workers since 1904 (Haldane), who reported that air temperature itself caused minor impairment, but that humidity caused significant physiological stress. While later research has confirmed air temperature as causing discomfort irrespective of humidity, it is generally agreed that high humidity conditions are particularly exhausting for those engaged in physical work. The Health and Safety Executive (United Kingdom) proposes that humidity between 30% and 70% does not impact comfort. However, particularly above 80%, high humidity reduces the evaporation of sweat and so, impairs our ability to cool ourselves. Anyone who has spent time in a Midwestern summer can likely testify to this! Prolonged exposure to these conditions, if not managed correctly, can lead to dehydration, fatigue, cramping, or at the extremes, heat stroke. The Occupational Health and Safety Administration (OSHA) proposes use of the heat index, a single metric incorporating both relative humidity (%) and absolute temperature (°F), as a guideline for risk to workers (Fig. 3.30). Put simply, the heat index suggests the perception and effect of a given temperature on the human body. Working in direct sunlight can add up to 15 °F to the heat index, and strenuous labor or wearing of heavy, protective clothing can also have adverse effects. An indication of the risks of humidity-related disorders is shown in Fig. 3.30 (https://www.weather.gov/safety/heat-index). Heat advisories and warnings may be issued by your local or national meteorological agencies. If you are working in areas where humidity may be high, pay close attention to these announcements. The Occupational Health and Safety Administration provides detailed advice on working in humidity which you should consult, but general precautions

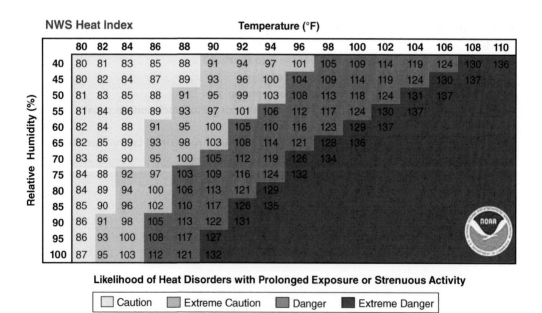

Fig. 3.30 Heat index risk ranking (Occupational Health and Safety Administration). *Source:* From Heat index risk ranking, National Weather Service. National Oceanic and Atmospheric Administration. US Department of Commerce. Public Domain.

include supplying adequate drinking water, planning for emergency, wear sunscreen, schedule frequent, sheltered breaks, monitor yourself and your teammates, schedule fieldwork for times when risk is lower, wear wicking, cool clothing, and always cease fieldwork if the risk is too high and adequate precautions cannot be implemented (Figs. 3.31 and 3.32).

It is important to acclimatize to conditions over time; heat-related workplace injuries occur most frequently in workers who are relatively new to those conditions. Acclimatization involves gradually increasing time exposed over 7–14 d; this may take longer in older or physically unfit workers. The National Institute for Occupational Safety and Health (NIOSH, 2016) advises that a little-and-often approach is best; 1 cup (8 oz) of water should be consumed every 20 min, while the Department of Defense suggests that in very hot conditions, during strenuous labor up to 40 oz (1 quart or 0.95 L) may be consumed per hour. Regarding clothing, light, breathable fabrics are ideal in humid conditions; however, you must still be protected from other hazards, so boots, gloves, masks, and other protective gear should not be omitted.

Fig. 3.31 This researcher is repairing irrigation equipment in humid conditions. A wide brimmed hat and sunscrean will help offset the effects of these conditions, but be aware that high humidity can amplify the discomfort experienced during hot weather. *Source:* Bo Collins.

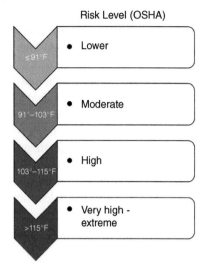

Risk Level (OSHA)

- Lower (≤91°F)
- Moderate (91°–103°F)
- High (103°–115°F)
- Very high - extreme (>115°F)

Fig. 3.32 Risk of heat-related disorders (https://www.weather.gov/safety/heat-index)

At the other end of the scale, cold temperatures can similarly impair your comfort and performance, although through different processes. The risk of workplace accidents can be elevated due to reduced dexterity as a result of cold, clumsy hands or heavy gloves (Anttonen et al., 2009) and possibly due to rushing or distraction. Prolonged exposure to very low temperatures can lead injuries of various severity including chilblains, frostnip, frostbite, and hypothermia. Your extremities (fingers, toes, ears, nose) are more vulnerable to cold injuries due to lower blood flow, low fat, and conservation of heat to the critical internal organs. These areas should be protected, and be alert to changes in color, numbness, burning sensation, or in extreme cases, blistering. These are all indicators of frostnip (milder harm) or frostbite (severe harm occurring when tissue cools to below 0 °C). Chilblains present as swollen and itchy extremities (toes and feet), which can often be discolored blue or red. They occur when the tiny blood vessels (capillaries) become inflamed as a result of exposure to cold or rapid changes in temperature (such as moving from the outdoors into a warm building or vehicle). They can be distracting, uncomfortable, and even painful. The best approach is to avoid them by wearing warm, dry socks and shoes and avoiding overly tight clothes which restrict circulation.

Hypothermia occurs when body temperature drops to below 33 °C (normal temperature is 37 °C). The body will feel cold and pain which may be accompanied by shivering and skin discoloration particularly of the lips and exposed areas. Increasing numbness occurs as the decline continues and the person may surprisingly become unaware of the severity of their condition. If someone is suffering from hypothermia, provide blankets or extra clothing, ensure that they are dry, provide calorie-rich food, and contact emergency assistance. In extremely cold conditions, awareness of your team members is essential as people are often have poor awareness of their symptoms.

Appropriate clothing is key to working in cold conditions. Multiple layers are recommended as opposed to a single heavy item, although in extreme conditions, snowsuits may be required. Warm hats are essential, and gloves and scarves may also be helpful. Mittens or shooting/fishing gloves which allow more complete dexterity will help if you have delicate work to do. Don't forget that you are vulnerable to UV exposure even in cold conditions so sunscreen may still be required. If you are not moving around much your temperature is more likely to drop. For example, downloading from a datalogger might require some time spent relatively still as you connect and interact with the computer. This cooling can be exacerbated by standing on frozen or bare ground. Placing a rubber mat beneath your feet can slow this as it provides a layer of insulation between you and the ground. Be aware that increased energy expenditure due to cold can make you hungrier than usual, so carry additional food; ideally something energy rich such as chocolate, biscuits, trail mix, etc. Be careful when moving from very cold to warm conditions, as rapid warming of your extremities can be extremely painful.

Low humidity can also be uncomfortable. Cold air is less capable of holding moisture, causing it to feel "dry." In these conditions skin, especially lips and sinuses can become dry or chapped. Again, sufficient hydration is essential, and moisturizers and balms may help ease damaged skin. Remember, chapped skin is more vulnerable to damage and cracking during physical labor.

Wind

I have only a very brief comment on wind in fieldwork; high wind makes note taking and sampling more difficult! Sheets blow away, sample containers spill, it is uncomfortable and can increase the perceived chill. In windy conditions, you ought to wear extra clothing as though the air temperature were colder than the dry bulb temperature. Consider using a ruggedized laptop if available or use a shelter when writing.

Sunlight

When planning fieldwork bear in mind that you may be constrained by visibility or daylight hours. Of course, if you are conducting nocturnal research, a bat survey for example, this may not be an issue. However, for the most part, fieldwork requires at least visibility and as this declines the risk of accident or error is likely to increase. Both season and latitude will influence daylight hours, so plan accordingly.

The Centre for Disease Control recommends avoiding direct sunlight during midday hours, wearing a brimmed hat, wearing clothing which covers your arms and legs, using sunscreen with an appropriate sun protection factor (SPF), and wearing sunglasses with UV protection. Keep in mind when working on or near water that glare can be particularly hard on your eyes. Sunscreen loses its potency during storage so throw bottles away within two years. Sunburn can be debilitating but is often undetectable until >4h after exposure. Symptoms range from tender, reddened skin, headache and fatigue, to nausea, blistering and peeling of skin, dehydration, and eye damage. If afflicted, drink water regularly to rehydrate, treat burns with topical moisturizers or hydrocortisone and lightly bandage if necessary. The effects of sunburn are often underestimated and although they typically pass within several days, the long-term effects of exposure can be severe. The risk of skin cancer as a result of prolonged exposure to ultraviolet radiation in sunlight has been increasingly recognized. Cancer Research UK reports that 1 in 36 UK males and 1 in 47 UK females are diagnosed with melanoma skin cancer in their lifetime and that the vast majority of these cases are preventable (Cancer Research UK, 2018). Both melanoma and non-melanoma forms of skin cancer are associated with exposure to UV rays in sunlight. Unlike many of the hazards you will encounter in the field which have an acute or immediate effect, skin cancer may not become apparent until many years after your exposure. Consider also that over the course of your long and successful career in the field, you are likely to accrue many hours of exposure. Essentially, you ought to follow the same preventative strategies as for sunburn; cover your skin, use a suitable sunscreen, avoid prolonged exposure, etc. Critically, be conscious of any changes to your skin such as development of asymmetrical discolored patches or lumps, growth of moles or freckles or areas of broken skin (with no obvious reason). If you are concerned, consult your medical doctor as early as possible.

Soil Conditions

Although soil conditions are not "weather" per se, they do fluctuate over time and are directly influenced by antecedent weather. Soil temperature and moisture are often key conditions of interest. Not only do these vary temporally but also spatially across the landscape and through the soil profile.

Temperature will strongly influence crop growth, microbial activity, and other biogeochemical dynamics, but precluding frozen soils is relatively unlikely to change the trafficability or workability of the soil. In other words, it may affect your measurements or samples, but in most cases, it is unlikely to strongly influence your ability to take those measurements. Moisture is perhaps an even more influential parameter as it effects biogeochemistry, biology, solute transport, gas dynamics; it fluctuates quickly over time and varies greatly across area. Not only will it likely influence your measurements and samples but also the trafficability of soil (its capacity to be traversed by a vehicle or on foot) and its workability (how it responds to digging, manipulating, sampling, etc.) will affect your ability to sample and even access your site. A PhD student I worked with installed pore-water sampling arrays in four sites; at one site, soil conditions were perfect for digging and working with – not too dry and hard to dig, not overly wet and sloppy. At that site, he and his team completed installation in under 5 h. At the other end of the spectrum, one site was so damp and unworkable that it was too challenging to install the samplers without risking irreversible structural damage to the plots. At that site, it took three days to complete the installation! The drastic difference was entirely down to soil workability.

Many national and local meteorological agencies supply both soil temperature and moisture data and often forecasts also. This data is useful for planning, however, be aware of the scale at which it is recorded, as soil characteristics are so spatially variable. I generally rely on these data as a guideline for directional change in parameters, but where possible, installation of monitoring devices at your sites is far more reliable and indicative. Temperature, moisture, and matric potential probes are increasingly inexpensive and available, and dataloggers can be equipped with telemetry for remote communication. Environmental monitoring is discussed further in Chapter 5. Regarding forecasts of soil moisture, these are generally made via numerical models and often take the form of a percentile (with 100% indicating full or effective saturation of soil pores) or a deficit relative to saturation (i.e., indicating mm or inches of moisture required to saturate the soil). These forecasts are excellent tools for guiding your work, but again, be cognizant of their spatial limitations.

River Conditions

Watercourses are innately hazardous environments to work in. Hazards include drowning, damp conditions, and zoonoses. However, watercourses are a key element of the hydrologic system, are habitats for a vast diversity of plant and animal species, are intricately involved in nutrient cycling and contaminant transport and supply much of the water we rely on. Water research is, therefore, an important discipline which strongly depends on field studies. As with soil conditions, river conditions are not exactly "weather" but are strongly influenced by it and may similarly affect your fieldwork.

A hydrograph measures the flow of a river at a specific location over time. This requires knowledge of the cross-sectional area of the river channel at that point and measurement of the "stage" or height of the water above a fixed datum, usually the riverbed. This indicates the volume of water passing through that location. The hydrograph can be disentangled to indicate the proportion of water arriving via groundwater, overland flow across the soil surface and interflow which comes laterally through the soil profile. According to weather and the nature of the watershed, the river will be influenced to a greater or lesser degree by each of these pathways. The amount of water flowing through the channel is known as "discharge," and it is indicated by the letter "Q."

$$Q = VA$$

Where V is velocity and A is cross-sectional area. Velocity is measured using a current meter. This device is a propeller which is rotated by the flowing water. The number of revolutions in a unit of time indicates the velocity, and by plotting multiple readings velocity against stage you can create a discharge curve. The discharge curve then allows you to record Q over time simply by monitoring river stage. Stage is the height of the water above a fixed datum. Stage can be measured visually using a staff guage or can be recorded ongoingly using a pressure sensor connected to a datalogger.

Q10 is discharge in the 90th percentile (in other words, which is higher than that which is observed 90% of the time) and generally classed as "high flow." This typically occurs as a result of a storm event and often is significant from a pollution perspective as runoff from critical source areas can occur. Fieldwork during high flows has several implications:

- Increased risks when working in or near the river due to water depth and speed, moving debris, possible impaired visibility, damp conditions, etc.
- Greater preparedness is required if sampling during high flows as these are triggered by major precipitation events and so may be relatively sudden. To capture data during high flow events you must be ready to deploy to the field at short notice.
- Potential damage or malfunction of monitoring equipment. In-stream devices and outlet monitoring kiosks may lose power if mains lines are damaged, may themselves be swept away or damaged by falling or moving debris or may become clogged by sediment. You can program your dataloggers to send an alert to your phone or email if a malfunction occurs, provided there is reliable phone signal in the area.

Q70 (30th percentile or discharge which is lower than is observed 70% of the time) is often considered to be low-flow, although Q80 to Q90 are also seen in the literature. During these conditions, most of the water in the watercourse will originate from baseflow, that is, groundwater. It is important to differentiate between low- and baseflow however, as they do not necessarily correlate exactly. This will vary depending on the nature of the watershed.

"Flashiness" is the rate at which water level changes or responds to precipitation within the watershed. A flashy river will rapidly rise in response to rainfall. This may be because the watershed is very steep or has clay-rich, impermeable soils which generate a large amount of runoff, which travels quickly over the surface into the watercourse. Watersheds which have karst geology (highly porous limestone with subsurface tunnels allowing preferential water flow) may also be flashy, although the pathway which the water takes to the river is completely different. Flashy rivers may be more hazardous to work in than rivers with low flashiness, as water level and the speed of flow can fluctuate sharply. Keep this in mind if event sampling.

Concluding Remarks

This chapter has discussed the primary issues which you need to think about when planning your fieldwork; site selection, risk assessment, equipment, maps and diagrams, photography, PPE, vehicles and weather issues. These could be considered as the base or foundation of your plans. In other words, you might reflect on these issues first, before moving on to the more detailed or day-to-day logistics, which we will discuss in the following chapter.

Take the time to think; have I reviewed the where's, why's, how's, and when's involved in my field plan?

References

Anttonen, H., Pekkarinen, A. and Niskanen, J. (2009). Safety at work in cold environments and prevention of cold stress. *Industrial Health* 47(3), 254–261. doi:https://doi.org/10.2486/indhealth.47.254

Cancer Research UK. (2018). Cancer research statistics. Research UK. https://www.cancerresearchuk.org/health-professional/cancer-statistics/statistics-by-cancer-type/melanoma-skin-cancer(Accessed 19 Aug. 2019).

Fealy, R.M., Buckley, C., Mechan, S., Mellan, A., Mellander, P.E., Shortle, G., Wall, D. and Jordan, P. (2010). The Irish Agricultural Catchments Programme: Catchment selection using spatial multi-criteria decision analysis. *Soil Use and Management* 26(3), 225–236. doi:https://doi.org/10.1111/j.1475-2743.2010.00291.x

Healey, M., Jenkins, A., Leach, J. and Roberts, C. (2001). *Issues in Providing Learning Support for Students Undertaking Fieldwork and Related Activities*. Cheltenham: Geography Discipline Network.

Joselyn, S. and Savelli, S. (2010). Communicating forecast uncertainty: Public perception of weather forecast uncertainty. *Meteorological Applications* 17(2), 180–195.

Logsdon, S., Clay, D., Moore, D., and Tsegaye, T. (eds). (2008). *Soil Science Step by Step Field Analysis*. Madison, WI: SSSA.

National Institute for Occupational Safety and Health. (2016). *Occupational Exposure for Heat and Hot Environments*. Washington, DC: Center for Disease Control. https://www.cdc.gov/niosh/docs/2016-106/pdfs/2016-106.pdf (Accessed 21 Aug. 2019).

Ortiz, O., Porta, J. and Rodriguez, C.D.A. (2014). Criteria and recommendations for capturing and presenting soil profile images in order to create a database of soil images. *Spanish Journal of Soil Science* 4, 112–126.

Schoeneberger, P.J., Wysocki, D.A., Benham, E.C. and Soil Survey Staff. (2012). *Field Book for Describing and Sampling Soils. Version 3.0*. Lincoln, NE: National Resources Conservation Service, National Soil Survey Center.

Vakulin, A., Baulk, S.D., Catcheside, P.G., Antic, N.A., van den Huevel, C.J., Dorrian, J. and McEvoy, R.D. (2011). Driving simulator performance remains impaired in patients with severe OSA after CPAP treatment. *Journal of Clinical Sleep Medicine* 15(8), 246–253.

Welsh, K.E., France, D., Whalley, W.B. and Park, J.R. (2012). Geotagging photographs in student fieldwork. *Journal of Geography in Higher Education* 36(3), 469–480.

4

Logistics

Logistical challenges are something in which environmental researchers rarely receive any direct training, but it is a critical aspect to effective fieldwork. Logistics are the elements of organization and management that ensure the smooth execution of your planned experiment.

Communication

Communication is an essential workplace skill, and this is especially true in fieldwork. Simply put, communication is the exchange of information between groups or individuals. This makes it an essential part of field logistics, as a researcher will often need to coordinate or liaise with several different groups. Communication is a two-way system – it is *not* about what *you* say. It is about what *your listener* understands. It is the foundation on which all your logistical concerns depend. In many cases, in the field you will be working in close geographic proximity (e.g., on the same plots, in the same soil pit). However, if you are working outside of speaking distance or sight of one another, you should have a plan for how you will communicate. Mobile phones are ubiquitous now, but be aware, in remote locations you may not have coverage. Two-way radios are an alternative, and you should also expect to have a "last alternative." By this I mean, a prearranged time and location to meet if communications fail utterly. Think about the communication tools and methods available to you and which will be most effective in your scenario.

Face-to-face – Verbal communication allows you to gauge the response or understanding of your listener, to clarify uncertainties, and to address any questions (Fig. 4.1). It may not be possible to physically meet with project partners in all cases depending on time constraints and geographic location. However, video conferencing software can allow this limitation to be overcome. Face-to-face meetings may be particularly important prior to fieldwork, to allow your plans to be discussed in detail. It is best practice to follow up major project meetings with an email including the notes or minutes so that the conclusions are recorded for future reference.

Email – This is a crucial and ubiquitous communication tool. Bear in mind that the quantity of email communication can be overwhelming for some researchers. Keep your emails concise (don't risk falling into the "too long – didn't read" category!). It is helpful to make the purpose of the email clear in the subject line (e.g., "Fieldwork schedule for November 2021"). You can also set the priority of the email so that highly important messages are flagged to the recipient. While this is not limited to fieldwork, it is always important to check that you have emailed the

Fieldwork Ready: An Introductory Guide to Field Research for Agriculture, Environment, and Soil Scientists,
First Edition. Sara E. Vero.
© 2021 American Society of Agronomy, Inc., Crop Science Society of America, Inc., and Soil Science Society of America, Inc. Published 2021 by John Wiley & Sons, Inc.
doi:10.2134/fieldwork.c4

Fig. 4.1 Face to face meetings give team members the opportuity to discuss the plan in detail and allow the team leader to gauge their understanding. These pedologists are holding a morning briefing prior to their excursion. *Source:* Jaclyn Fiola.

correct address. Although this is not a comprehensive text on email etiquette, it should be highlighted that interpreting the tone of an email can be challenging. Items which are high priority can be underemphasized while comments meant casually can seem harsh or critical. The tone with which the recipient reads the email can be very different to that intended. Always read back your emails and evaluate how they might be understood by their recipient.

Phone – Mobile phone use has become ubiquitous, with 81% of U.S. adults owning a smartphone (PEW Research, 2019). These smartphones exceed many of the capabilities of traditional phones, combining email, text or web messaging, and social media. Phone calls have obvious utility in the field. Keep in mind the battery requirements, particularly if you are also using camera or GPS functions of the device, as these can be particularly power demanding. Also consider your data allowance and whether you will be charged at roaming rates. Make certain you have the phone numbers of everyone you might need to contact before embarking on fieldwork and distribute your own contact details. Consider the weather conditions and terrain also. In wet weather or if you are working near water, your phone could easily become damaged. You might consider investing in a waterproof, sturdy phone case. It will certainly be cheaper than replacing the phone!

Text messaging – Text or web-messaging is useful for short communication and particularly for sending photographs or brief video footage. While this is not a strict rule, I avoid using this method for very important information because texts can be misinterpreted if you are not clear and you may not be certain if the recipient has seen it (although some apps do indicate this). Particularly useful are the "group chat" functions of several messaging apps, as these can allow the entire fieldwork team to engage in the same conversation. If using one of these apps, make sure all members of the team have access to that software.

Radio – Phone signal may not be available in particularly remote locations or in terrain which has obstructed lines of sight. Citizen band (CB) radio or walkie-talkies may be a useful alternative. If using this equipment, you should practice beforehand to make sure you can use them reliably, as these tools tend to be less popular than other devices. Consider the power requirements and whether you will need an antenna. You should determine beforehand which frequency you will transmit on, and your callsign. Never use any frequencies reserved for private use, the military, or emergency services.

Social media – There are a wide variety of social media outlets which can be useful for communicating your research to the public, including Twitter, Facebook, and Instagram. "As it happens" reporting of your fieldwork can generate considerable interest in your research and make it accessible to stakeholders and the public who might not otherwise be aware of it. Be responsible in your use of social media. Avoid posting anything that could compromise your work, or that might have data protection implications (such as identifying a host landowner for example). Take care not to waste time on social media during fieldwork. It may be a useful addition, but it should not distract from the primary objective. Social media also provides potential employers and collaborators with an impression of you based on your online presence. Separating professional and personal media may be advisable (Lima, 2018). The ASA, CSSA, and SSSA host several sessions during the annual meetings on the use of social media by researchers. If you intend to use some of these outlets, consider engaging with your professional societies for guidance and to help develop your network.

Your Team

Within each team are different roles and responsibilities. The specific tasks that each individual is responsible for will differ depending on your project. Typically, the team leader will be responsible for organizing meetings, assigning tasks, devising the schedule, ensuring appropriate training and equipment is available, and making major decisions. Team members will be responsible for fulfilling their assigned tasks, complying with all health and safety regulations, and communicating with the leader and with other team members. It is essential that everyone understands who the leader is and who has the ultimate responsibility for decision-making. Conflict can occur when this role is either not clear or not filled. If the leader does not decide and effectively communicate to the team the likelihood of accidents or errors can increase. As the leader, you should listen to your team but take responsibility and be decisive. If you are not the leader, you should be a willing and constructive team player. A good team player is considerate of other team members and actively contributes to achieving the overall project goals.

Advice for being a good team member

- Be a good communicator. Tell your team leader what you see and learn. Communicate your needs and ask for instruction or clarification as necessary

- Follow instructions, but not mindlessly. Do not put yourself, others, or the overall objectives at risk because you were "following instructions!"

- Be constructive. Offer your knowledge as relates to the experiment. Try not to be a complainer! If you have a legitimate grievance or concern you should discuss it in a calm, clear manner with the appropriate individual and seek a solution

- Be punctual and reliable. Fieldwork is demanding and needless delays will waste valuable time and diminish your effectiveness. If you say you will do a certain task, see to it that you do so or communicate to your team leader if you need assistance

- Be patient with other team-members

- Be flexible. Fieldwork often involves dynamic conditions, genuine hazards, physical discomfort and limited timeframes. If you can adapt to these and maintain your performance, you will be a valuable member of the field team

- Be diligent and conscientious. See that you do all your tasks to the best of your ability, even if it is not your own experiment. The success of a team or a project depends on all contributors fulfilling their duties

- Look out for your team-mate's safety. This includes checking that everyone has received the necessary training and equipment ahead of time, and that safety protocols and good communication are maintained during fieldwork

---

Your team needs to know and understand the overall plan, their individual roles/tasks, how to use their equipment, and what the schedule is. A preparatory meeting in the days leading up to fieldwork allows these issues, and others, to be discussed. It is a good idea to provide a written procedure or schedule for team members to follow. Establish a clear line of communication. The team should know who to ask for guidance, who is responsible for each task, and what to do if something goes wrong. You can demonstrate techniques and discuss how potential scenarios should be handled in the field. Make certain that each individual knows when and where you will meet, what they should wear or bring, and what is expected of them on the day. Even if you have worked together before, it is good practice to refresh and revise. You should provide risk assessments in advance of fieldwork and give your team the opportunity to ask questions, voice concerns, or otherwise contribute usefully to the plan.

Ask your team what they have understood from your meeting. The most important question you can ask is "What did you understand from those instructions?" and have your team mate tell you what they believe their tasks or instructions to be. You might be surprised that their interpretation was different from your intention. Don't say "Do you understand?" They might believe that they do, when this is not the case! Equally, if you are the person receiving instructions, recite them back to your team leader so that they can confirm or correct as needed.

The division of tasks and responsibilities should be clear. Each team member should know what they need to do and how to do it (Fig. 4.2). Assignments should also be realistic. For

Fig. 4.2 My team investigating a soil pit! Fieldwork is a great opportunity for team building. *Source:* Sara Vero.

example, you should not set a target of eight samples per hour to any individual if it takes 10 minutes to correctly take each sample. This puts unrealistic pressure on your team and makes success impossible. Tasks should also be fair. Teams often struggle when the division of labor or responsibilities is perceived to be unfair. It may help to rotate tasks. For example, if Person A is driving to the site and taking herbage samples on Day 1, on Day 2, Person B might take a turn at these tasks, while Person A takes over downloading the data-loggers. Of course, specific skills can limit this, and you should always recognize and praise the contributions of each individual.

Fieldwork is a wonderful opportunity for team building and close friendships are often forged when you wade up rivers, dig pits, haul equipment, take long drives, and eat together. You will overcome challenges with your team, celebrate successes, see beautiful locations, and learn new things. However, there are two issues which you should consider.

First, even in the field you must conduct yourself in an appropriate manner to any workplace. Your institution likely has dignity-at-work or similar guidelines you can consult. Do not place yourself or any member of your team in situations which they may find compromising or inappropriate.

Second, keep in mind that you will become physically and mentally tired while doing fieldwork. No one performs or behaves at their best in such circumstances and people may become withdrawn, grumpy, or argumentative, even if you have formed close friendships and are an effective, efficient team with superb communication skills. I recall two team members heatedly disputing which direction a cloud was traveling after several extremely long and challenging days of fieldwork and travel. This silly dispute was laughed about later, but at the time, these two good friends were in sharp disagreement. Monitor your own mood and that of your team and never allow behavior to degenerate. I suggest taking regular breaks and also acknowledging when you are fatigued. It is completely reasonable to take a break apart from your team to focus. It is not reasonable to be rude, mean, or unpleasant to your team mates. Keep in mind, they are likely as tired and frustrated as you.

The Local Community

In some cases, you may need to communicate with residents and authorities in the area where you will work. Clear communication will ensure good relations and help your research to be successful. This is for several reasons:

Reasons to communicate with the local community

- Your work may be influenced or impaired by other peoples' activity
- Your work may present some hazard or disturbance to locals
- You may need permission to access an area
- People may be curious
- You may need co-operation or interaction

If you are accessing privately owned land you must ask permission in advance and let them know what to expect. Be aware that you may not be granted permission, in which case you must always respect the wishes of the landowner. Where a field-site has been established and both you and the other party understand that visits will be ongoing it may not be necessary to alert them every time. However, it is always courteous to let landowners know that you will be or have been on their property. Make sure that you have a clear agreement and always be respectful, considerate, and responsible. Remember, you *do not* have a right to enter private property.

Always have a means of identifying yourself and your organization. This can be clothing (e.g., high-visibility or logos) badges, I.D. cards, or papers. This can reassure people who may be alerted by or curious about your presence. This is particularly true if you are conducting surveys, question-naires, or assessments that require interpersonal interaction. Don't expect someone to give you information without knowing who you are. High-visibility clothing is helpful not only for your own safety but also to let people know that you are in the area. Work vehicles often carry logos or insignia but if not, magnetic I.D. can used.

While it is rare, negative interactions with the public do occur. Equipment can be vandalized or stolen, and I have heard several stories of dangerous encounters when working in the field. It is never worthwhile to put yourself or others in danger and you must always obey the law. If threatened or intimidated by a third party, you should remove yourself from the situation rather than challenge them. Incidents should be reported to your institution and, if appropriate, to the authorities.

Officials

Certain research may need permission, licenses, or access to be granted by government or other officials. These needs should be assessed well in advance of your intended fieldwork, as the pro-cessing of applications can be time-consuming and will likely be out of your control. It is essential to contact the relevant authorities at the earliest opportunity and, if possible, obtain an estimate of when the relevant documentation will be available. Always be polite and respectful when dealing with authorities and comply with official requirements to process your applications promptly. Remember, they may not be aware of your time constraints and may possibly have limited control over the processing times. Some common approvals include approval to conduct research on human or animal subjects, using regulated materials travel documentation, and permits for ship-ping plants and animals. Regulated materials may include potentially infectious agents, biological specimens, and select agents and toxins listed by the USDA (HHS and USDA Select Agents and Toxins, 7CFR Part 331, 9 CFR Part 121, and 42 CFR Part 73). These are substance deemed to pose "a severe threat to both human and animal health, or to animal and plant products." The exact requirements for research involving regulated materials varies depending on the material, the nature of your research, and on your location. Consult your research office when planning so that they can assist you with approvals and guidance.

Visas are critical when traveling to foreign countries. For many countries, a specific category of visa is designated for individuals traveling for research purposes (e.g., J-1, H-1 or O-1 visas for researchers coming to the United States). You should consult with the embassy or government website of your intended host country as to:

o The category of visa you need
o The application procedure
o The document/I.D. requirements
o The estimated processing times
o Permits, quarantine or other requirements pertaining to either sampling or transport of samples across borders

This final point is of major importance considering the threat posed to human health and ecology by invasive species and zoonoses. Never take this for granted or attempt to circumvent laws or procedures either deliberately or inadvertently. There may be rules pertaining to the collection and transport of both living or dead flora and fauna, geological and archaeological specimens, and soil and water samples. You must investigate this prior to initiating fieldwork.

The Buddy System

Buddy systems are popular in many fields including scuba diving, firefighting, mountaineering, education, and in the military (Fig. 4.3). While the specifics of each systems vary, in general, a buddy system is a pair of individuals deployed together who work in close proximity or communication. In the field, you may work as partners assembling equipment or taking measurements. Alternatively, you may use a more remote or distant system in which you communicate continuously or at agreed times. This is particularly useful in survey work where there are large areas to cover. Several ways in which a buddy system can be used are briefly described:

Training – Cooper and Wright (2014) describe partnering new employees with more experienced staff so that they can be mentored and informally tutored. This can be an effective way to acclimatize new team members to fieldwork, as besides the specific skills (i.e., how to take a soil core), a vast spectrum of knowledge can be accumulated with experience, such as labeling samples, dressing for the field, and packing your vehicle so that equipment doesn't spill everywhere as you try to pull out a critical tool! These aren't necessarily complicated techniques, but they are the difference between a smooth, successful day in the field and a stressful, slow and ineffective one. These skills may not necessarily be taught during university, but a training buddy can get a new team member up to speed (although I hope this book will help also!). Training a team-member is a considerable responsibility and the senior buddy must be willing to take part, have excellent communication skills and patience, and have experience with the environment, tools,

Fig. 4.3 The buddy system can be used in the field for efficient data collection or to help one partner learn new skills. These researchers are using a buddy system while collecting soil respiration measurements on Konza prairie. *Source:* Jesse Nippert.

and methods they are demonstrating. If you are the senior buddy, remember that venturing into the field for the first time can be quite daunting for new researchers, who may be more familiar with controlled environments, regular hours, and ready access to equipment, spares, and restrooms! You may need to guide your buddy in things that you now take for granted. Be patient. Think back to your own early days in the field!

Cross-checking – Buddies or partners can oversee the preparation and correct use of safety equipment and supervise measurements taken by the other person. This approach is very common in diving, firefighting and other dangerous pursuits and professions, but can be used to safeguard against mistakes, errors, and accidents in your research.

Touchstone – When you are out of range of verbal or visual communication a pre-arranged schedule for checking-in can limit accidents and allow action to be taken should a check-in be missed. I operated this system with a friend who was conducting stream sediment sampling at a remote site. My friend was operating alone, but we had arranged for phone calls at pre-set times throughout the day. If she did not ring in at the scheduled time, I was to alert nearby technicians who could rapidly deploy to the site. This is known as a "*dead-man's switch*" system. In other words, action is triggered by a lack of human response, rather than an alert system. This is ideal for potentially dangerous situations such as working alone in water or in remote areas, as accidents may render the victim unable to issue an alarm or call for help if they are incapacitated.

Clean hands/dirty hands – This is a system of teamwork proposed by USGS (2006) for water quality sampling; however, it is ideal for almost any type of environmental sampling where contamination could compromise sample quality. In this system, one team-mate (clean hands – CH) conducts all tasks which involve direct contact with the sample (e.g., handles the sampling bottle or bag, takes a water, soil or herbage sample, places equipment inside any sterile chamber or container). The other team-mate (dirty hands – DH) conducts any tasks which may involve contact with sources of contamination (e.g., handles the exterior of equipment and containers, writes observations and completes field-sheets, drives vehicles and operates machinery or equipment used at multiple sites). Both team mates must adhere to good gloving and sample hygiene procedures, but this approach adds an additional level of security which might be particularly useful when sampling microorganisms or volatile compounds, where small amounts of contamination can make a sample useless for research.

Keys to a successful buddy system

These tips will help you to be a part of a successful buddy team

- Get to know your partner
- Listen to them
- ...and communicate back
- Be considerate
- Anticipate their needs
- Be patient
- Offer help and use their assistance

Checklists

You will have realized by now that when venturing into the field you have many things to consider, plan, and prepare. It is not reasonable to expect that you will remember absolutely everything unless you have a strategic approach. The best way to do this is by making checklists. You can keep lists of (i) things to do and (ii) things to bring. I can't tell you what should go on your lists – they are unique to what you are doing, where you are, and what you need. However, as an example, here are my "To Do" and "To Bring" lists from a fieldtrip to install some soil monitoring arrays (Table 4.1).

Table 4.1 Example check-list for installing a monitoring array.

To do	To bring	
Check weather	Boots	Cool-box
Contact landowner	Hat	Bottled water
Book accommodation	Waterproofs	Lunch
Talk to team	Notebook	First aid kit
Print maps	Bulk density rings	Data-logger
Pack and refuel van	Mallet	Cables
Ring soil scientist	Sample bags	Solar panel
Talk to lab staff	Augur	Sensors x 12
	Screwdriver kit	PVC pipe
	Shovel	Multitool
	Duct tape	Sunscreen
	Tarp	Clipboard
	Camera	Permanent markers
	Toothbrush/toothpaste	Bucket
	Toilet roll	

The list is divided into things to do prior to deployment and items to bring.

The Importance of the "Trial Run"

It is good practice to always conduct a trial run before beginning fieldwork, particularly where new techniques or use of unfamiliar equipment is involved. You will not want to discover that a certain apparatus is not functioning when you are already on-site, or that it simply takes longer than expected. You can conduct your trial run somewhere conveniently local to your research station. Practice until you are confident that you can use and understand your equipment and note how long it takes to perform a measurement. If you are already familiar with the procedure you may not need additional practice, however, you should at least check that all devices, tools, and equipment are functioning correctly. Don't consider these trials and checks to be wasted time. You will waste far greater time and resources by deploying to the field with malfunctioning gear or without enough experience.

In the Field

The day of fieldwork can be a stressful one, although thorough planning ahead of time can reduce stress. Also, collecting, checking, and loading your equipment the previous day can save time on the day of fieldwork itself. Fieldtrips can be single or multiday events, local or distant, physically demanding or low intensity. However, here is a simple plan for ensuring your fieldwork is successful.

Before Fieldwork
1) Check the weather forecast.
2) Contact site owner/manager or other involved outside party. Remind them that you will be on-site, what time and any other issues that they should be aware of.
3) Coordinate with laboratory to receive or process samples.
4) Gather all tools or equipment and check that they are working correctly.
5) Gather and check personal protective equipment (PPE).
6) Print maps and diagrams.
7) Fuel and load vehicle.
8) Pack food and water.
9) Brief team and arrange travel, including planning route and rest stops.

During Fieldwork
1) Travel to site.
2) Check in with team and confirm that everyone has arrived on-site and knows their role.
3) Conduct measurements and/or treatments.
4) If working separately or remotely, check in with team members or contacts at agreed times.
5) Take regular breaks, including food and water.
6) Label and pack samples. Ensure that they are chilled if necessary and stored securely.
7) If you install sensors or other devices, check that they are logging or transmitting.
8) If attending in situ equipment such as monitoring arrays, check their batteries, inspect for damage or wear, and perform any required calibration, etc.
9) Collect and repack all equipment, gear, and clothing.

After Fieldwork
1) Deliver samples to laboratory or appropriate storage on time
2) Unload vehicle and leave in clean condition for next individual
3) Copy or download notes, data and photos. Back up your information!

Things Go Wrong!

Despite planning and preparation, things can and do go wrong during fieldwork. It is your job to anticipate potential setbacks, to have a plan to deal with them, and to be adaptable and practical when dealing with the unexpected. The ways in which things go wrong in the field can be divided into delays, errors, accidents, and vandalism, theft, and interference. Each of them will be discussed in this section.

Delays

Delays are very common in fieldwork and can occur for a variety of reasons (Table 4.2).

In some cases, delays may be out of your control. However, effective and thorough planning can prevent many of these. If you allow sufficient time plus some extra you can often achieve your goal within a realistic schedule. In research, it is vital to be realistic rather than optimistic or naive when estimating how long it will take. If you are leading a team, you must recognize that members may operate at different speeds. Some people are intrinsically motivated and driven in the field, irrespective of whether it is their own research or others'. Some people are less so. Similarly, some people are indeed motivated but due to personality, physicality, or disposition work at a very slow, but steady pace. In short, do not be surprised or irritated if all members of your team do not match your pace and you may need to factor additional time in light of this. You may need to adjust your expectations. Ask yourself "Is the speed with which I expect this task to be done both *reasonable* and *realistic*?" Do not ask a team member to operate at speeds which are beyond their capacity.

Weather is a common cause of delays. In some instances, particular weather conditions are essential (e.g., event sampling of overland flow). There is little you can do to expedite this; you may need to be "on standby" so that you can act once conditions are suitable. This is discussed in greater detail in the "Weather" section. Weather can also cause delays if it makes traveling safely to your site slower or if it causes you to need more time in the field. Safety is crucial. If you need to drive in inclement weather (rain, snow, ice, or high winds), you will need to schedule additional time and may need to postpone completely if conditions are dangerous. In the field, additional time may allow you to continue sampling if conditions are mild (e.g., a light rain) and if you are equipped with suitable clothing. However, if conditions are hazardous (e.g., a thunderstorm), you may need to abandon sampling for that day and seek shelter. There is no strict rule of thumb for

Table 4.2 Common causes of fieldwork delays.

Prefieldwork delays	During-fieldwork delays
Ordering and delivery of equipment and consumables	Late or slow team members
Institutional approval	Bad traffic
Suitable weather conditions	Heavy rainfall
Availability of the site or personnel	Sampling taking longer than expected

this; you must decide based on your assessment of the risk. It is better to accept a delay than to risk the safety of you, your team mates or a third party.

Errors

An error is an incorrect, inaccurate, or compromised measurement or result. Errors can generally be divided into three categories: systematic, random, and transcription.

Systematic Errors are mistakes that are consistent and directional. This creates *bias* in the measurements. An example of this might be if a balance is calibrated incorrectly, and so consistently reports weights 2% higher than is actually true. Systematic errors are problematic; however, if the direction and magnitude of the error can be quantified you may be able to salvage those measurements by adjusting to the correct levels.

Random Errors are mistakes that are inconsistent in direction and/or magnitude. This creates *variability* in the measurements. An example of this might be if the balance is correctly calibrated but is not placed on a secure, level, and stable base. Consequently, when you are weighing samples it both over- and underestimates measurement in a non-consistent fashion. Keeping to the balance example, another random error could occur if the sample was contaminated by a heavier material (such as a rock mixed in with a herbage sample). This would lead to a heavier weight being recorded for that sample than is true, although the other samples in the set would be correct. Random errors are more problematic than systematic errors, as it is difficult to pinpoint both cause and the severity. Furthermore, as it is not directional (in other words, both an under- and an over-measurement can occur in the same set of measurements), there is no single correction factor which can be applied.

Transcription Errors occur when a measurement is incorrectly recorded or when duplication or download of the record from field notes to your files is incorrect. There errors can occur very easily, particularly if you do not take steps to write clearly, legibly, and consistently. Make sure your field sheets have enough blank space to record your observations in large, neat writing. Carry spare pens and pencils and keep paper dry if possible. A voice recorder can be a useful tool for recording observations. I have used one when describing soil pits, and then transcribed my observations onto pit cards back at the office. If you are working with a team, you should emphasize the importance of taking good notes. If you cannot read the measurement from the field sheet, then that data may essentially be lost from the study. Make certain that everyone uses the same abbreviations and acronyms. It is best to duplicate or download measurements and notes as soon as possible after fieldwork, as the measurements will be fresh in your memory, and you are more likely to correctly record them. Don't assume you will remember details in a week, especially if you are taking many measurements or observations.

It is good practice to scan or copy field sheets once you return and to download or backup data, notes, or photographs saved on electronic devices including phones, tablets, laptops, and GPS devices. Both notes and devices can easily get lost or damaged. Loss of data in these instances is not only frustrating but also a waste of your time and resources.

Standard Operating Procedures and Quality Control

One way to avoid errors and omissions in the field is to consistently follow the same sampling procedures and orders. Routine will help you to adhere to standard operating procedures (SOPs). If you have several actions to take at a particular site (i.e., record location, measure stream parameters, take a grab sample, and fill in a field-sheet), you should always conduct these tasks in the same order at each sampling point (USGS, 2006).

An SOP is a written document that provides detailed instructions for a routine procedure. They are ubiquitous in laboratory work but are sometimes (incorrectly) overlooked in field research. Standard operating procedures are essential to maintain consistent treatments and measurements, particularly if several individuals are using the methods, when training new team members, and when writing materials and methods sections in papers or theses. They also help avoid the problem of "insider knowledge" – in other words, when only one person in the research group knows just how to use, find, or maintain certain equipment or methods and hence, all work depends wholly on them. If that person is unavailable, whole research plans can grind to a halt. The SOPs must be clear, concise, and detailed so that they can be successfully applied by the reader. A poorly written SOP may only confuse details. Follow the style used by your institution so that readers will know what to expect and where in the document to find the relevant information. The standard operating procedures should be available in hard copy and also stored electronically if your institute operates a shared drive. In general, an SOP will have five parts: the title page, table of contents, procedures and methods, quality control and assurance, and references (EPA, 2007). The title page should include name of the writer and the date of preparation, along with dates of any updates or revisions. Systematic review every 1–2 years of all SOPs used by your research group will help you keep up-to-date with new methods and approaches. The EPA recommend that every SOP be issued with a short identification code to maintain good archiving and allow documents to be easily located. It is best practice for each SOP to detail a single procedure or method. If your planned fieldwork involves several discrete tasks (e.g., taking soil bulk density cores, applying a dye tracer, and harvesting vegetation using a quadrat and shears) you should prepare an SOP for each individual task. If the SOP is long, a table of contents may be helpful, although this might not be essential for simple or brief methods. The procedures or methods section should include the following:

- A brief summary of the method.
- A list of definitions. This is critical if jargon, abbreviations, or acronyms are used that could not reasonably be expected to be understood by relatively unfamiliar readers.
- Qualifications or training an operative will need to perform the method or procedure.
- Risk assessment and health and safety protocols.
- Equipment and supplies including everything you will need to conduct the method. Be detailed. The reader will need to know the specifications of any equipment; for example, the volume of glassware, type of gloves, size of filter paper, etc.
- Step-by-step instructions on how to complete the procedure. These instructions should be simple, concise, and specific. They should include instrument specification and calibration, sample collection, handling and preservation, methods of analysis, data recording methods and approaches for troubleshooting.

Methods of quality control should be recorded next; these are discussed in detail later in this section. Finally, you should include all references used in the preparation of the SOP. This will guide the reader as to the development of the procedure and allow them to develop a deeper understanding, if they require it. Once you have written an SOP, it should be reviewed by a sufficiently experienced individual so that any errors or omissions can be corrected. If you are a PhD student, for example, your supervisor might be a suitable person to help with this.

Quality control (QC) includes systems and procedures used to detect, avoid, and/or correct errors and is a crucial part in ensuring your data is suitable for analysis and publication. Quality control procedures must be defined within the SOP and then applied to all measurements taken in the field, just as is the case for laboratory measurements. For chemical analyses, a "blank" can be used to detect systematic error. In this approach, a sample of distilled or deionized water or a premade solution of known concentration will be analyzed using the same procedure and equipment as the field samples. This will indicate how accurately the device is measuring. You can then calibrate your probe, sensor, or device as per the manufacturer's instructions. Similarly, balances can be checked using calibrated weights, and ovens using temperature sensors. Consult the manuals for your equipment for the appropriate QC and calibration procedures. Continuously monitored data such as stream discharge or air temperature must also be checked. It can be challenging to keep up to date with this, particularly with high-frequency monitoring. Essentially, the entire dataset must be visually assessed for (i) unrealistically high or low measurements without explanation, (ii) "flat-lined" measurements that do not fluctuate at all, or (iii) blank spaces in the dataset indicating either a lack of measurement or a failure to log it (Fig. 4.4). Depending on the type, cause, and extent of errors in monitoring data, you have three basic options: (i) discard the compromised portion of the dataset, (ii) interpolate (estimate a value based on known data) the missing data based on trends, past records from equivalent monitoring periods, or statistically calculated values, or (iii) integrate values from nearby or comparable measuring stations. A good example of this is precipitation, wherein many catchment or watershed studies will have a secondary rain gauge that provides a source of backup data.

You should always record the date, time, methodology, and results of your QC checks. If you are using handheld probes or sensors, field balances, or any calibrated equipment or devices you should check their accuracy before going to the field, and at scheduled throughout the year. This should be done for equipment in storage also, especially if you may need to deploy that gear at short notice, such as for event sampling.

Accidents

We can think about accidents in two broad categories: (i) those pertaining to people and (ii) those pertaining to an experiment or equipment. Regarding people, following risk assessments and implementing safe working practices should minimize the likelihood of harm to your team. Health and safety legislation typically require that workplace or work-related accidents must be reported within a specified time frame to a designated individual. This law applies even if you are working in the field. Find out the procedure for reporting to the responsible person in your organization. Depending on the nature of the accident, you may also be required to report to an outside authority, such as a local fire chief, police, Centre for Disease Control (CDC), health and safety authority, or others. You must document precisely what happened when doing so as there may be legal or insurance implications (determining liability and remediation) following such incidents that might require accurate records.

Accidents involving an experiment or equipment may be caused by you, a third party, or simply an uncontrollable occurrence such as flooding due to extreme rainfall or a burst pipe, animal damage, loss of samples, damage during transport, or other such events. There is no predefined solution to any accident, as the necessary actions will be specific to your experiment. Before you begin fieldwork, find out whether your university or research institute has specific instructions or a required sequence of responses which you should take following an accident. As a general approach, in the event of an experimental or equipment accident you should:

1) **Stop and Think**. This is the most important step. When something goes wrong it is tempting to leap into action. However, as with any part of research you should have a clear, logical reason for whatever you do, and it should be conducted according to a sound methodology. Acting in haste rarely achieves this and may compound the damage done already. Do not overreact.

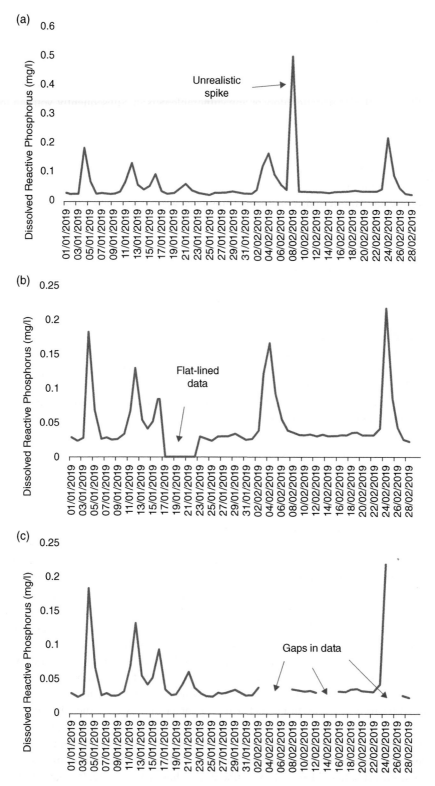

Fig. 4.4 Data recording issues; (a) unrealistic values, (b) flat-lined data and (c) gaps in data. *Source:* Sara Vero.

Keeping calm and thinking about possible solutions will keep things in the correct perspective.

2) **Ensure the Safety of You, Your Team, and Third Parties**. This may mean first aid, contacting emergency services, leaving a dangerous or compromised environment, and/or raising an alarm. The specific order of these will depend on the situation. You should prioritize whichever action will secure the immediate safety of any vulnerable person. Subsequently, you can follow up with further action as appropriate. Emergency services should be contacted when "a person's life, health, property, or the environment is in danger." Emergency service operators may ask you the address of the emergency, directions, landmarks, and details on the incident including who is involved and what has happened. When speaking with emergency service operators by phone you should stay calm and follow their instructions.

3) **Define What has Happened**. When something goes wrong, it is easy to misunderstand the scenario. You should clearly and accurately describe what happened and why. This allows you to make informed decisions about what to do next.

4) **Report to the Correct Authority**. Reporting of accidents and near-misses is a requirement under health and safety legislation in many countries. Follow the reporting procedures and chain of command specified by your institution. This will allow the situation to be reviewed and any legal requirements to be fulfilled. Keep in mind, if a law has been broken the correct authority to which you should report may be outside of your workplace, for example, the police.

5) **Determine the Implications**. Depending on the nature and magnitude of the accident, the implications for you and the study will vary. You need to figure out whether this incident has compromised the experiment or puts you at risk. The following questions will help:

   - Can new samples be taken? Consider cost, time, and resources.
   - Will new samples be reflective of the treatment or has the accident impaired or influenced the experiment beyond the initial damage or loss?
   - Can the equipment be repaired?
   - Can the experiment be adapted?
   - Can the experiment be restarted?
   - Should the experiment be abandoned?
   - Are the results publishable with the caveat that the accident is noted?

6) **Determine and Implement a Solution**. Based on the information and your analysis of the incident you must decide your solution or response and implement it in a timely fashion. Make sure that your collaborators agree with your decision. You should involve them in your solution-finding process. Repeating the experiment is not necessarily the optimum solution. It may be that your design or the circumstances are simply not suitable or cannot provide the information you require. In such cases an entirely new or intrinsically different experiment may be required. The appropriate and effective solution to any accident will of course depend entirely on what exactly happened and what the implications are. However, like conducting a risk assessment, you must follow up this mental activity with practical action. This may take some time, as you may need replacement equipment or additional help. It is important when implementing a solution not to be hasty, but rather, to do what will be most effective.

7) **Get Over It!** It can be very upsetting and frustrating for a carefully constructed experiment and much hard work to fail. If this is the case and you cannot rescue your efforts, it may be essential to simply acknowledge the loss, learn from it, and move on. Failed experiments are generally not published but that certainly does not mean that they don't happen. Ask around and you will find many stories of both field and lab experiments that have gone wrong. It is part of a research career and an opportunity to learn and to develop new skills.

Vandalism, Theft, and Interference

Although it is always disappointing, vandalism of scientific equipment is not uncommon. It may be done in opposition to the research in question or simply a random act. While data are not available, anecdotally it seems that visibility is the biggest factor in vandalism. Vandalism can be relatively minor, such as graffiti, which may not impair the functioning of the equipment, but sometimes can completely compromise measurements or destroy the device. Replacing damaged or destroyed equipment can be extremely costly.

Theft is typically and thankfully rare; most of the public have very little use for research equipment. One item is at higher risk – batteries. Data-loggers and other equipment frequently use 12- or 24-V batteries that may be connected to solar panels. These are potentially vulnerable to theft.

Interference is usually not malicious, but may be no less damaging to the experiment. People are often curious when some activity or installation occurs that is new to the environment and rarely realize their presence may be disruptive or even harmful. Try to be patient when dealing with the public and explain politely but firmly that their interaction is detrimental to your work. Most people will accept this and behave accordingly. As discussed in the communications section, having identification and appropriate permission for your activity can be helpful in this regard.

Making equipment less visible is likely to prevent damage, theft, or interference. This is not always possible or desirable. Signage may help; in which the ownership and purpose of the apparatus is indicated, and contact numbers provided should a third-party need or want to interact. Communicate in advance with local people. Security cameras may be useful; however, you must make yourself aware of the legal requirements and data protection rules pertaining to such devices. Unfortunately, simply recording theft or vandalism does not prevent it from occurring. It may be sensible to equip logger-boxes or kiosks with secure locks to discourage theft.

Moving On

Try not to be disheartened by setbacks but rather deal with them and be positive. Talk to your colleagues; everyone has encountered challenges. I have cancelled fieldwork due to drought, broken equipment at critical moments, fallen into rivers, experienced vandalism, had samplers gnawed by wild animals, and had a logger-box torn apart by a hedge-trimmer. Remember, it is not the end of the world when things go wrong. My PhD supervisor gave me great advice; he said that research teaches you to be a problem-solver, so these experiences just make you a more useful person. Similarly, I think these are useful sentiments for a field researcher, although I am sure you will have many more victories than defeats. Your future employers may or may not care about your specific scientific findings or skills; however, your ability to adapt and overcome challenges will be invaluable in any situation.

Taking Care of Yourself in the Field

During fieldwork you have dozens of things to remember, tasks to accomplish, concerns, objectives, and deadlines. It is very easy amid all of this to overlook the one critical and irreplaceable component: you. Even assuming no accidents or errors, there are many ways that your performance can be undermined or that fieldwork can become unpleasant, difficult, and exhausting. However, some very simple practices can prevent this.

Before we delve into the various aspects of fieldwork self-care, there is one point which stands out from all others. *Always let someone know in advance where you are going and what time you will return.* Accidents can and do happen and none of us are invulnerable. If no one knows where you are your chances of rescue in case of disaster are at best delayed and at worst, completely impossible. Time is a key element in survival. If no one knows when you expected to return, they are unlikely to realize if something has potentially gone wrong. Often, the difference between a near-miss and irreparable harm can be the speed of rescue. If you need help because a logger breaks down, a vehicle is trapped in mud, or some other practical inconvenience, your contact must be able to find you. Consider the stress and fear that can occur for family, friends, and co-workers if you or a member of your team is "missing-in-action" or in worse cases, physically harmed. Accidents have far-reaching impacts on the lives of those close to the injured individual including financial burdens, changes to household tasks, psychological stress, and emotional upset (Dembe, 2001).

Dressing for the Field

Being correctly dressed has a major influence on the success of a fieldwork trip. There is a Swedish truism that there is no such thing as bad weather, only bad clothing. In general, it is best to wear multiple layers that you can remove and replace as necessary.

For an average day, working on land in mild weather, the basic fieldwork outfit should be as follows:

o Underwear. It might seem silly to offer advice on which underpants to wear but mild discomfort under typical conditions can become extremely distracting during fieldwork. Wear what is comfortable, not what looks best! Sports-brand underwear is often ideal as it is designed for high levels of activity, sweat and movement. For women, chose a comfortable, well-fitting bra. Depending on your build, a sports-bra or other non-wired bra might be preferable as wires can chafe and become uncomfortable. The field is not the time to wear something for the very first time, as you don't want to discover that it is uncomfortable, poorly-fitting, or requires adjustment.

o Always carry at least one spare set of socks. Socks are one of the most important elements of my fieldwork kit. Wet feet will blister, tear, and sweat and are vulnerable to chilblains. Damage to your feet can incapacitate you from fieldwork and is considered to be of such importance that modern armies provide explicit training for their troops on correct foot care. I stick to a fixed routine; once I have finished fieldwork for the day (or if my feet become wet) I remove my wet socks, completely dry my feet, apply talcum powder to absorb any moisture and put on fresh, dry socks. If my boots are damp, I will change them also. I always follow this routine before getting in my vehicle to drive. Be wary of developing blisters; these can burst and may be vulnerable to infection. There is an elevated risk of infection if you are working in watercourses. Chilblains can be remarkably uncomfortable and distracting. Keeping your feet warm and dry are the best prevention. Sudocream can be used to treat them should they develop, and vitamin B supplementation is believed to limit their occurrence. I can't stress this enough; *look after your feet*! Just as you should try several pairs of boots, try out different socks to go with them and be aware; not all "non-slip" socks are truly that. Typically, socks made from blends with a higher Lycra content are less prone to slipping. Personally, I find a pair of light sports socks (which go to the ankle) with a heavy pair of hiking socks over them to be very effective.

o Footwear should be selected depending on your task for the day – the same boots may not be suitable for all situations. If the ground is not too wet and if you have large distances to cover, hiking boots are a good option. You can use high or low boots as you prefer, although high-ankle boots offer greater support, which may be helpful if the terrain is rough or steep. You should select non-slip soles. When shopping for hiking boots be sure to try several varieties as they can differ significantly in comfort. In my experience, it is worthwhile investing in good boots as they will last several years. If you are using machinery of any sort (e.g., motorized soil corers, weed-eaters/strimmers, mowers, etc.) steel-toed work boots are recommended for safety. These boots are reinforced to protect your toes from impacts, cuts, and falling equipment. If you choose steel-toed boots, you should be aware that if you are walking long distances, they can be more tiring to wear due to their weight and may be uncomfortable as your toes rub against the reinforced cap. Safety should always be your first concern, so if it is necessary to change footwear during the day, do so. Wet or muddy conditions call for waterproof boots or wellingtons (nicknamed for the Alfred Wellesley, first Duke of Wellington), which are often also available with reinforced caps. Avoid wellingtons which have overly wide legs; these will flap as you walk and are vulnerable to becoming filled with water. Be careful if you step into streams or ditches of unknown depth; a wellington full of water makes for a very unpleasant day! Something to note with wellingtons; socks are highly prone to slipping down and becoming bunched at the ankle or mid-foot. This is not a major problem, but it is inconvenient to hitch them up repeatedly. *Never* wear a new pair of footwear for the first time in the field. This is a recipe for disaster as new boots almost always require some "breaking in" as they adjust and soften to the shape of your feet. Un-broken shoes will chafe and can cause blisters.

o You may bring a pair of shoes or sneakers to change into after fieldwork. Not only will this be more comfortable when driving or traveling but also, allow you to enter shops or restaurants without bringing mud, water, or worse into their premises. Remember, just because you don't mind the dirt and grime of the outdoors doesn't mean others feel the same!

o Full-length trousers. Hiking trousers often have multiple pockets and straps, which can be convenient for stashing equipment. In addition, they are frequently available in moisture-wicking or quick-drying fabrics. Denim jeans perform poorly in wet conditions. The capillary action of that fabric causes it to retain moisture and for damp to creep upward from your ankles. Be aware of this is you are walking through grass or vegetation early in the morning as dew can saturate your lower legs remarkably quickly. Reinforced work-trousers such as those worn by builders and trades-people are excellent for most fieldwork. The fabric is strong and resilient and unlikely to tear easily. In addition, the knees are frequently padded or can be fit with reinforced inserts offering additional protection. I strongly recommend trying these on before you purchase as sizing can differ greatly and if you are of a relatively small build it can take a little effort to find well-fitting trousers in this style.

o Long-sleeved shirt or base layer. This will offer both warmth and protection against sunlight and insects. In cold conditions thermal base layers are advisable. These are available in several weights and thermal ratings.

o Pullover or sweater. A lightweight, insulating sweater provides the next layer. Fleeces or microfiber tops can be effective and are thin enough to allow jackets or other layers to be added.

o Vest (sleeveless sweater). I wear this over either my base layer or pullover, depending on air temperature. These keep your core warm, without being smothering. There are many options including fleece, down, or waterproof versions. "Fisherman-style" vests can be useful as these have pockets and tags for equipment.

o Bring a spare set of clean, dry clothes. Clothing often becomes damp, dirty, and sweaty during fieldwork. It is often a great relief to change into fresh clothing before driving home. I recall spreading cattle manure on study plots with a friend of mine. Neither of us had brought a spare set of clothing and during our 4-h drive home we did not feel comfortable stopping in a coffee shop that we passed on account of our rather bad smell! We should have been better prepared. Even when you are not engaged in fieldwork, which is quite that messy, you can be rained on, spill something, or rip clothing so it is sensible to have a backup.

o Your head is particularly vulnerable to exposure to temperature extremes, sunlight, and wind. You may need a variety of hats. Brimmed or peaked hats (like baseball caps) are not only useful on hot days, but in any bright weather (Fig. 4.5). While the capacity for heat loss through the head relative to the rest of the body is disputed, it is typically a highly exposed area. If you are in cold environments, you should wear a comfortable, well-fitting hat. Ear-flaps will help protect vulnerable extremities. Avoid headwear which obscures your vision. Additionally, reflection from snow or water is extremely tiring to your eyes, as is prolonged driving. Comfortable UV-protective sunglasses such as those used by golfers, rowers, and hikers are recommended.

Fig. 4.5 Your head is vulnerable to the cold and to sun. These researchers wear baseball caps while soil judging. Reseachers need to wear many hats, both figuratively and literally! *Source:* Karen Vaughan.

○ Neck-sleeves or face masks (such as Buffs) are useful, for additional warmth in cold weather and for protecting your mouth and nose if moving through dense vegetation. They can be conveniently folded and stashed in a coat or bag pocket.

○ Wet-gear. There are many options for water-proof or-resistant clothing. As regards to rain jackets, heavy jackets offer the greatest level of protection against the elements; however, if the air temperature is not cold you may sweat or become overheated. I generally keep my heavy jacket in my vehicle for when I need it, but don't typically wear it often. A light jacket is often preferable as this can be put on as a final layer over your thermal or other clothing. It is preferable for raincoats to extend below your hip, ideally to mid-thigh. If possible, sleeves that have Velcro or button fasteners are preferable, as these can be fastened tightly about your wrists. This prevents water running up your arms and keeps heat in.

Light, waterproof overtrousers are helpful in rain or snow, and when working in soil pits, around animals, or in any other environment that is damp or mucky. These can lose their water-resistance over time, so either maintain by treating with waterproof coating or replace as necessary.

Working in rivers brings different demands. A lifejacket must always be available if you are working in or near watercourses. Regarding clothing, wellingtons and waterproof trousers may be sufficient if the watercourse is very shallow (e.g., an open ditch or small stream). However, waders that incorporate boots, trousers and a waterproof bib all in one piece are preferable when working in any waterbody which rises above your ankle (Figs. 4.6 and 4.7). You can purchase waders in hunting or fishing or outdoors stores. You must still wear trousers and a shirt under these waders.

Fig. 4.6 Waders incorporate boots, waterproof trousers and a bib. They should be used when working in water up to waist-deep. *Source:* Katherine O'Reilly.

Fig. 4.7 Working in rivers requires specialist clothing such as waders. Check that your field clothes are suitable both for the environment and your physique. *Source:* Krista Keels.

Always be cautious as if water spills over the top of the bib it can fill the waders and may make it extremely challenging to get out of the watercourse. Be aware that this may raise your risk of drowning. Some waders are designed with quick-release straps so that they can be rapidly abandoned if you become submerged.

It is possible to purchase outdoor clothing that is pre-treated with permethrin or other insect repellents. These should be considered if you are working in areas which have ticks, mosquitos, or other biting or stinging insects (Fig. 4.8).

Fig. 4.8 Biting or stinging insects such as mosquitos can be extremely irritating and worse, may spread disease. Use PPE such as long sleeved clothing and head nets, and insect repelant as necessary. *Source:* Jaclyn Fiola.

Hydration

Dehydration is a common cause of impaired performance. In the absence of sufficient intake to replace lost fluids, sweating reduces blood plasma volume and consequently impairs muscle and brain function. This reduces productivity and increases the risk of workplace accidents. Symptoms of dehydration include reduced urination, dizziness, headache, reduced skin "springiness" and in more extreme or chronic cases; weakened pulse, loss of consciousness, or kidney damage. Loss of over 1% bodyweight as fluid loss represents a threshold indicator of dehydration, with serious health implications occurring as that percentage increases. A simple indicator of hydration is urine color with darker shades usually indicating poorer hydration status (Fig. 4.9).

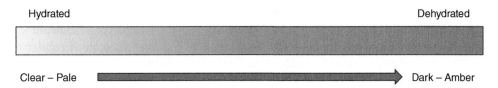

Fig. 4.9 Urine colour is an indicator of hydration status. *Source:* Bo Collins.

The first rule of hydration is to do it before you need it. If you wait to drink until you are thirsty, then you are already somewhat dehydrated. This does not mean you are in danger, but it does mean that it will be challenging to fully rehydrate as you go about your work. Interestingly, workers have been found to maintain hydration status over the course of their shift, not improving from an under-hydrated state despite fluid intake (Miller and Bates, 2007). This suggests that establishing initial good hydration is critical, as it may be difficult to take in enough liquid during strenuous labor, particularly in hot conditions. Miller and Bates (2010) reported chronic and prevalent underhydration in outdoor workers and a 2007 study by the same authors found 51% of outdoor workers ($n = 710$) displayed hydration levels which put them at increased risk of poor performance and workplace accidents.

The National Institute for Occupational Safety and Health recommends an intake of one cup (8 oz) of fluid every 20 min, while Miller and Bates suggested 600 mL to 1 L (c. 20–34 oz). However, in warm weather and during physically demanding tasks you will need more than this. Water should always be available, although electrolyte drinks may be useful. Be aware of what you are drinking as it is easy to rely heavily on sugary or carbonated drinks when on the road. You should encourage a "culture of hydration" among your team (Miller and Bates, 2010). In other words, don't assume or leave it to chance that you and your group will hydrate; you must but in place protocols and tools to ensure it. That study suggested setting minimum fluid intake goals, supplying each team member with their own 2 L bottle, monitoring hydration, and placing reminders to drink in visible areas. During fieldwork, consider writing **REMEMBER TO HYDRATE** in the margin of field sheets, on your vehicle dashboard, or set alarms on your phone at regular intervals.

A nice habit I learned during fieldwork in Kansas: freeze several plastic bottles of water the night before fieldwork. These slowly defrost during the day and so remain cool and refreshing and will simultaneously keep your lunch cool.

Food

It is essential to bring enough food on fieldwork, particularly if you are somewhere remote or when leaving the site to purchase food is not feasible. Hard physical work is likely to make you hungrier than usual, and you should bring foods which are filling, energy-rich, and easy to digest. Fieldwork is not the time to experiment with new foods. Bring a lunchbox or cooler and plenty of snacks in addition to your main meals. Trail mix, fruit, protein bars, and other packaged foods are ideal, although personally I prefer to avoid very salty foods such as chips/crisps. It is best to avoid foods that can spill or leak, or those that might be prone to spoilage in heat or if packaged for too long. You may also want to avoid foods that need to be heated as these facilities might not be available to you. If you bring salads, carry your dressing in a separate small container to prevent leaves becoming soggy and wilted. It's a good idea to carry a set of cutlery and napkins.

While snacks can carry you through in a pinch, it is wise to take at least a lunch break at an appropriate time and to rest fully and eat. Sharing food with your team is also an opportunity to build relationships and encourage team cohesion. I was fortunate to be part of a fieldwork team in which the leader frequently brought homemade granola bars and maintained regular rest breaks. This showed great care and consideration for the team and was a morale boost when we were tired.

Something I never leave at home is a toothbrush and toothpaste. This might seem unimportant, but a lingering taste can become increasingly unpleasant, not to mention the importance of good hygiene. Brushing your teeth is unlikely to be of critical importance to the success of your experiment however, small actions like this have big consequences for your degree of comfort in the field, which in turn influences morale and performance. Hygiene in relation to eating is crucial. Always de-glove, wash your hands, and use a disinfectant spray or gel before eating. Don't run the risk of illness due to laziness or lack of preparedness.

Toilet

On several occasions' students and even quite experienced researchers have told me that they avoid or are uncomfortable urinating or defecating while on fieldwork and remarkably, may limit food and drink because of this. This is an extremely bad strategy. It is uncomfortable, will slow you down, and will make you less efficient in your work. Dehydration is discussed in detail above, and anxiousness about toilet access is no justification for this. Female researchers in particular often find the lack of facilities in the field challenging, although that is certainly not always the case. Regardless of who you are, there are a few simple principles that can help and, as with most aspects of fieldwork, planning, and preparation is the cornerstone for success.

First, always use a real toilet when the opportunity presents (e.g., at gas stations, rest stops, if you are in a restaurant) – even if you don't think you *need* it at that time.

Second, always carry small toiletries on your person. Miniature tissues, wet-wipes, etc. will make it easier and more comfortable to look after yourself in the field. It is best to keep these items in a zipped pocket on your clothing. If you keep them in your backpack or other storage, it is easy to forget them when you need them. Always bag and properly dispose of used tissues and paper products just as you would if camping or hiking. If you are operating at a site for a number of days, particularly if also camping there, dig a small hole for a field-toilet and replace soil when you leave. The University of Sydney recommends that a hole 20 cm deep located at least 100 m from the nearest watercourse.

Third, don't wait until you desperately need the bathroom. As with eating, drinking, and resting, during fieldwork, you are better off using the toilet (or a convenient bush) when you need to, not when it has become an emergency.

Finally, don't feel nervous or embarrassed. You can simply excuse yourself, move away from your group to a sheltered location, and do what is necessary. Remember, it is completely natural and something that everyone on your team will need to do also. Set a good example by treating this issue in a mature and practical fashion and work toward removing any stigma around this issue.

On a practical note, urination in the field is generally a lot simpler for men. However, for women and for anyone who needs to poop outdoors there is the matter of how. Simply squatting is one option, however, in the western world ceramic toilets are ubiquitous and many people aren't familiar with the practice. If you lack the leg strength for squatting, an excellent alternative is to lean your back or shoulders against a solid prop such as a tree or rock, keeping your feet well in front. This is much easier for most people and generally fairly comfortable. There are a number of female urination devices (FUDs) now available which allow women to urinate standing up. This can be an advantage in the field, both for privacy and when removing multiple layers of clothing might be challenging. Reusable and disposable versions of these devices are available, and practice prior to going to the field is recommended to prevent spills. Female urination devices have become very popular with hikers, climbers, and other outdoorswomen and there are a wide variety of makes and styles now available.

There are a few potential hazards to keep in mind when toileting outdoors. You probably don't want to be disturbed as this might be embarrassing for you and for the accidental intruder. You should move away from trails or public areas, both for privacy and hygiene; however, be careful not to wander too far away, fall, or get lost, particularly in rough terrain or when light is dim. It is good to let a team member know that you are leaving so that they can be aware if you are missing for an overly long time and always stay within calling distance. Another concern is insect bites. These are bad enough in general, but in a sensitive area, they could be unbearable. If you are at an insect-prone site, use repellent sprays and avoid vegetated areas that are particularly prone to ticks, etc. Depending on what part of the world you are in, be aware of snakes and spiders. You should choose your temporary toilet location wisely; stinging or spikey plants such as nettles, poison ivy, brambles, etc., can make for a truly unpleasant experience. Cast an eye around before you go. A final practical concern, fieldwork clothes are often loose, have straps and tags, belts, etc. and you want to avoid soiling these. Try to pull any loose clothing items out of the way before you do your business.

Moving in the Outdoors

To experienced hikers, hunters, and anglers moving through the outdoors in rough terrain or farmland may come as second nature. For many of us however, there is a learning curve that can take many people unawares (Fig. 4.10). There is a technique to everything from crossing fences and streams to interacting with livestock.

If you need to cross a fence, ditch, or river you should do so at the first safe location. You do not know whether another safe crossing exists up or downstream and you may have to walk a significant distance out of your way searching for one. Take the opportunity where it presents.

Fig. 4.10 These researchers are dressed for a long day in the outdoors. Plan your clothing from head to toe. *Source:* Bo Collins.

If you are crossing particularly difficult terrain, long distances, or are carrying a pack, walking poles may help you bear the load and reduce fatigue. These are widely available from outdoor stores.

Always leave gates as you found them. Open gates should be left open, closed gates should always be shut again after you pass though. Likewise, locked or unlocked gates should be left as they are found. On farmland, gates are closed for a good reason! Leaving it open might be tempting so that you don't have to get out of your vehicle to reopen on your return, however you might be letting livestock or wild animals in or out where they are not wanted. This could ruin your experiment or put you and your team at risk. Furthermore, open gates can result in loss of stock or damage to crops. This will greatly impact the landowner and is unacceptable behavior for researchers hosted on private land. In such events you are unlikely to be welcomed back and may be liable for resulting damage.

If you need to climb over a gate, pass your equipment to the other side, rather than carry it over the gate. If you have a team-mate (see section on the Buddy System) with you one of you should cross first and the second partner should pass the equipment over. When climbing a gate always climb near to the hinge. This puts less pressure on the gate (think leverage!) and also, is less likely to swing or shake. Metal gates can be slippery. Always climb carefully and pass one leg at a time over the top. Never jump off a gate and never climb a spiked gate.

When parking your vehicle take care not to obstruct roadways, gates, or paths. This is inconsiderate of the landowners, local community, and of the general public. Furthermore, it can prevent emergency access or pose a hazard to other drivers. It is best practice to park in such a manner so that you do not need to reverse and can simply drive out directly. This is safer. If necessary, ask a team member to guide you in parking or leaving thereafter.

Electric fences and gates are a potential hazard on farmland, particularly in areas where cattle graze. Many farmers will post a warning sign, but this is not always the case (Fig. 4.11). Furthermore, it is all too easy to accidentally touch a wire even when you are aware that it bears a charge. I can personally attest to this! While clearing vegetation from an access point to a ditch I accidentally touched a fence that I had believed to be turned off. This was not the case, and the jolt from the fence flung me off my feet. I was thankfully unhurt but did feel unwell for a couple of days. The voltage of electric fences can vary greatly depending on the animals being enclosed or excluded. Voltage hurts, but it is the amps that actually cause injury. Most livestock fences have relatively low amps but can still deliver an extremely unpleasant shock. Despite this, don't hesitate to contact your doctor if concerned or unwell after receiving a shock. Never touch someone who has been shocked, as the current can pass to you. If you hear a snapping or clicking noise near a fence, it is certainly electrified. If you do not know whether a fence is on or not, assume that it is. Only open electric fences using the plastic handles at gateways (Fig. 4.12). Remember also that fences and gates can be poorly installed or earthed or may malfunction. In these cases, the severity of the shock delivered can be significantly more severe.

Barbed wire is common on agricultural land and can snag clothing or pierce your skin. You must also be careful not to damage fencing when attempting to cross. If there is sufficient space, it may be easier and safer to slide or roll under a wire fence rather than climb over it. If you absolutely must climb a fence, always do so as close as possible to a post as this will be the most stable location. Never climb or roll under an electric fence. You may become trapped and shocked.

If you need to get down a ditch or bank, find an access point that is not obstructed, slippery, or steep. It is often safer to slide down banks rather than attempting to climb them. Climbing can be treacherous as soil and vegetation can give way under your weight even when they appear relatively sturdy. I recall a team mate stepping on a branch while trying to climb down into a

Fig. 4.11 Not all fences will be clearly indicated like this one. *Source:* Sara Vero.

Fig. 4.12 Be careful around electric fences. These can give you a nasty or even harmful shock. Watch out for plastic handles like this for opening electric gates. *Source:* Sara Vero.

channelized streambed. The branch suddenly gave way, causing him to drop 3 m below and completely out of sight into the watercourse. He was luckily unharmed. A safer way to do this is to sit on the bank and slither down. Of course, this should only be done if your trousers are thick and sturdy so that they do not rip or snag and ideally, and when the vegetation is dry or your clothing is waterproof. You may be able to kick vegetation out of your way as you go. Take your time when moving down banks and ditches. Shield your eyes if possible as branches can strike or poke them, which at best is stingy and at worst can cause serious and lasting harm.

Fatigue

Fieldwork is frequently physically and mentally demanding, involving long hours, travel, and sometimes quite challenging labor. While an occasional day in the field is relatively easy to recover from (and often, a welcome break from the lab or office), prolonged field campaigns involving multiple days away from home take a toll. Furthermore, irregular hours that are either longer or simply different (earlier or later) from your typical schedule can reduce performance, motivation, and safety. A large-scale survey of US workers conducted by Rosekind et al. (2010) revealed that workers with irregular sleep schedules showed impaired time management and performance relative to those with regular schedules. More alarmingly, they experienced highly significant increases in accidents, near misses, and unintentionally nodding off both while at work and when driving. These safety impairments were even greater than those reported by workers suffering from insomnia. It is also worth noting that even aside from the additional physical fatigue accrued by physical labor in the field, it can be difficult to sleep the night before fieldwork, perhaps due to anxiety or excitement. Cumulative sleep debt is accrued over multiple 24-h periods with insufficient sleep to meet your individual physical demands. These can cause levels of fatigue on par with entire nights of wakefulness. There is no "quick-fix" solution for cumulative sleep debt; you need successive days involving at least 7 h sleep per 24-h period.

The aviation industry has identified the "window of circadian low" or "WOCL." This is the period between 2 a.m. and 6 a.m., for which most people acclimated to a typical day-night/wake-sleep schedule correlates to impaired performance and alertness. If you are operating during this period, particularly if you are not familiar with it, be particularly aware of potential errors and increased risk of accidents (Transportation Research Board and National Research Council, 2011). I have a personal experience of this. I was conducting river sampling, and on one occasion, samples needed to be taken early in the morning so that they could be delivered to a particular laboratory for analysis. The laboratory was located approximately 6 h away. I arrived on-site at the river at 5.30 a.m., and from the beginning I made errors that were uncharacteristic. First, I approached the river from the opposite bank from usual and as a result was less familiar with the slope. Accidentally, a sample bottle fell into the river and began to float away. What I should have done was fetch a net from the vehicle, move downstream and tried to fish it out. Instead, I instinctively stepped into the river to grab it. Coming from the unfamiliar bank, in dim lighting, and within the WOCL, I had made a dangerous error. The water was deeper than I had realized (not to mention cold!) and higher than my waist. Thankfully, I was wearing my life jacket and was able to climb back out, get dry and warm (thanks to my spare clothes), and continue with my day. I knew that I had been very fortunate – only small changes to that scenario could have led to harmful or even fatal results. These accidents can happen to anyone, and they happen in seconds. The WOCL does not apply only to pilots; be wary in the early hours.

While they cannot restore you to a completely rested state, naps are a useful tool for short-term fatigue management. Short sleep periods (minutes to a few hours) are advised by road safety agencies as a strategy to reduce driver fatigue and are increasingly instituted in workplaces to improve employee productivity. Research by Brooks and Lack (2006) evaluated the effects of 5, 10, 20, and 30-min naps, versus a no-nap control on the alertness, performance and "sleep inertia" of participants. Sleep inertia is the period of drowsiness, low body temperature, and disorientation which occurs after waking, the duration of which can vary depending on individual and conditions (Muzet et al., 1995). Brooks and Lack found that 10-min naps were overall the most effective, yielding improvements for up to 155 min thereafter. This improvement was offset in longer nap periods by more prolonged sleep inertia; in other words, people who slept longer had greater difficulty in becoming fully alert once they awoke. This 10-min duration is often recommended by road safety agencies for drivers undertaking long trips. If you are in the field, you may have opportunity to nap in a sheltered, safe area or in your vehicle; however, you should note that time napping in a semirecumbent position (e.g., in a car seat) does not fully equate to that equivalent length of time in a bed (Transportation Research Board and National Research Council, 2011). Nevertheless, it is a truly helpful strategy and something I utilize frequently, particularly if working in the field for several consecutive days. Make sure that you are in a safe location away from moving vehicles, machinery, or animals. You should aim for a comfortable temperature and adequate ventilation. Always set an alarm so that you do not oversleep and never sleep in an exposed area. If you are sharing a vehicle with your team, you can take turns driving while the other team members nap. Some people believe that they simply cannot nap; it is not part of their makeup. However, just as temperature can be acclimatized to and specific skills can be learned, so too can napping habits be practiced and developed over time (Muzet et al., 1995).

Evaluate both your sleep patterns and your level of fatigue over time. You will probably recognize obvious symptoms such as over-tiredness, drowsiness, nodding off, or decreased exertion. There are more subtle indicators too including crankiness, giddiness, or becoming over emotional. Interestingly, fatigue can impair your ability to moderate your body temperature and can cause you to feel far more chilled than weather conditions alone might lead you to expect.

Personal Safety

Aside from the hazards related to the environment or your activities, it should not be ignored that threats can arise from other individuals as well. If working alone, you may be particularly vulnerable, and while incidents are rare, they do occur. I know individuals who have been verbally abused, physically intimidated, and even targeted by vehicle thieves. As with all hazards, you should have a plan for how you will deal with personal threats like these, should they arise. This plan will resemble best practice for personal protection in your life outside of fieldwork, but bears remembering.

1) Always act on your intuition. If you feel that a situation is somehow "wrong," you must act. This sense was discussed in detail in the excellent book by Malcolm Gladwell entitled *Blink* (2005). That book is worth reading in full, but to summarize, your experience collated over a lifetime of interaction with other people allows you to interpret physical signals at a very subconscious level. While we might not be able to consciously pinpoint it, we can all feel threat or intimidation. If this occurs, act! Victims of crime often report that they "knew something was wrong" but did not act. Of course, this does not in any way infer blame toward them, but it does tell us that the first step in protecting ourselves is to be aware of our surroundings.

2) Be aware of the people around you. This is not limited to people who may wish you ill. Remember that third parties may be curious about what you are doing or startled and alarmed to come across someone unexpectedly, particularly in remote locations or on private property. People may react with hostility, so carry identification. Always be polite and respectful and move away from conflict immediately. If you are working on farms or in locations where machinery is operated, be aware that those individuals may be completely oblivious to your presence and could inadvertently pose a threat. Don't become so fixated on your task that you are not aware of others around you.

3) Be aware of your own behavior and presentation. Make yourself aware of local customs and attitudes prior to fieldwork in regions culturally different to your own and avoid causing offense or seeming "out-of-place." Even in areas with which you are familiar, appearing lost or distracted can leave you vulnerable to theft or threat. While personal interactions and cultural engagement are a rewarding part of fieldwork, be aware that political or controversial topics might be contentious and there may be subtleties of language, history, and opinion, of which you as an outsider are not aware. Approach such topics with caution.

4) Move away immediately. It is tempting in the face of any hazard to press on with your work or to assume that the risk is either low or worth taking. However, it is very difficult to assess risks when it comes to personal safety. If you are threatened or intimidated in any way or suspect that such a situation may arise, leave the area entirely and immediately. Do not return to a compromised or risky environment or situation. Just like you should not re-enter a building after a fire-alarm has sounded, so too should you keep away from any situation that has become unsafe.

5) Sound the alarm. If you are threatened, you need to get loud. You must attract attention and alert others to what is going on. If you can move away, go to a public location and tell someone what has happened. You may need to phone the police or other relevant authorities and you should inform your manager.

When operating in areas experiencing social or political unrest or where crime rates are high, you are likely to be at particular risk. Researchers have been kidnapped or caught amid civil unrest. If operating in such regions, you may need private security including drivers and vehicles. This should be discussed with your research institute in advance.

Stress

For many agricultural, environmental, and geophysical researchers and students, fieldwork is the most engaging and enjoyable aspect of their work. The outdoors is what inspires many individuals to enter these disciplines (Fig. 4.13). Speaking for myself, most of the fondest memories of my working life have taken place during sampling, installing equipment, wading up rivers, and even driving

Fig. 4.13 Fieldwork can be stressful. Don't forget to take a moment to appreciate the outdoors. *Source:* Sara Vero.

to and from field sites with people I liked. You can look forward to fieldwork and recognize the genuine opportunity for adventure that lies ahead of you. However, it would be naive to deny that fieldwork can be physically and mentally stressful. Proper preparedness as discussed in this manual should help limit stress, but it is important to remain aware of your own response to the challenges and pressure you will encounter. If you find yourself becoming angry, upset, overly worried, or in any other way emotional as a result of your fieldwork, you should keep things in perspective. Excessive or unrealistic workloads (even self-imposed) and lack of support or facilities are common stressors. Furthermore, stresses encountered in any workplace may be difficult to leave behind when you go home at night.

The American Psychological Association recommends keeping track of your thoughts and noting habitual or patterned behavior. This will help you to identify stress responses so that they can be addressed. Be alert to disordered thinking such as catastrophizing, fretting, and discounting successes. Watch your sleep and eating patterns and make efforts to "switch off" by socializing, resting, and maintaining nonwork relationships. Your research institute may have stress management programs, or you can talk to your supervisor if it is becoming an issue for you.

It may seem unlikely, but self-sabotage is not uncommon as a response to stressful situations. An individual may deliberately or subconsciously make mistakes to relieve pressure and release their obligations. For example, key equipment may be "forgotten" resulting in sampling being cancelled. I know this sounds unlikely, and genuine mistakes happen of course. However, if you are experiencing high stress, you should carefully and honestly examine your actions. Ask, are you doing something that is negatively affecting your work, and if so, why are you doing this? If you are honest, you should be able to differentiate the causes of your actions and whether they are truly informed by the situation or if they are a response to your emotions. In the latter case, these actions are unlikely to truly solve the problem and can only provide short-term relief. If you find yourself acting this way it may be helpful to talk to a team member or to seek outside help with whatever aspect of fieldwork is problematic. You may need to exercise self-discipline in your approach and keep in mind problems can be overcome by identifying and addressing them. Self-delusion on the other hand, cannot improve your situation and likely, will result in greater expenditure of resources and potentially can exacerbate the problem.

My PhD supervisor had an excellent approach that I have tried to remember: every day is new. In other words, if one day had been utterly disastrous, he did not bring that failure into our next day – it was over and done with, and we had a clean slate to move forward with our research. We did not dwell on it. On the other hand, if we had a big success or made significant progress, the next day I was still expected to step up, work hard, and move on to the next objective. In this way, our project constantly moved forward. This is a very positive attitude to bring to your fieldwork, particularly during challenging times. Even in challenging times, it is important to show appreciation for your team members' efforts. Remember to celebrate successes and forgive mistakes.

Try also to find the things that bring you a sense of satisfaction or achievement, whether these are large or small. Remember, every time that you successfully take a measurement, complete a survey, or maintain equipment you are advancing our shared understanding and developing your skills. A day in the field is rarely wasted. I have spent innumerable hours wracking my mind in the office trying to disentangle results, get some statistical code to work, or to bring a manuscript to completion. It is sometimes difficult to discern progress on one of these days. I can honestly say though that I have experienced very few days of fieldwork that did not yield something useful. Keep your eyes open.

Accommodation

If you are spending several days in the field, are distant from your research center or are traveling to multiple locations, you may need to stay overnight at or near to your field site. If you intend to start fieldwork very early in the day, or work late into the evening, it may also be helpful to have accommodation so that you do not have to travel overly early or late. I strongly recommend using overnight accommodation to minimize the amount of time spent driving in a single day, particularly in addition to the demands of fieldwork.

Hotels – A comfortable bed and sometimes even a swimming pool are certainly welcome after a tough day in the field. You can also arrange for early breakfast or other requirements, particularly at larger hotels. Be aware that hotels may be expensive; check your budget in advance. Most hotels will offer corporate rates that you may request. Some research agencies require hotels to be booked in advance via a specific booking agency or protocol. Make certain you check this with your administrator in advance or you may find it difficult to be reimbursed subsequently.

Hostels and B&Bs – Areas that may not have large or chain hotels may have a bed and breakfast or other alternative, although they may not offer the same amenities. Keep in mind that B&Bs are frequently operated from peoples' homes, so be considerate regarding noise, time of arrival, and particularly, cleanliness. Returning from the field, your clothing may be damp or soiled. Try to avoid making your temporary accommodation overly dirty (this applies for most accommodation).

Camping – If you are in a particularly remote location or conducting nocturnal research (e.g., bat surveys, monitoring live catch traps), it may be necessary to camp on or near your site. Many excellent books are available on camping, but specifically regarding fieldwork you should consider that this imposes additional requirements regarding equipment, time to prepare your tent, additional food, and suitable overnight clothing. Consider how you are going to transport this additional gear to the site. REI estimates that loaded pack weights typically range between 10 and 30 lb (REI, 2018).

Sample Management

Packaging Samples

You need to arrange the movement of your samples from their natural environment to a laboratory, while minimizing their degradation. How you contain your samples depends on the nature of the sample itself and on its size. Your container must be large and secure enough to contain the sample, should be possible to label, and may need to be sterile.

General Containers

Bottles, flasks, and vials – These are ideal for water or other liquid samples (urine, blood, milk, etc.) and can be purchased in various volumes and dimensions. Keep in mind that different plastics are used to form these containers, so check the manufacturer's specifications to ensure that they will not interact with your sample. Commonly, they will be sterile until first opened.

Plastic bags – These are suitable for loose samples (such as soil) provided that the samples are not sharp or overly heavy, which can cause bags to tear. Zip-top bags are particularly useful as they are easy to seal. Press-seal bags (which have a double, interlocking lip) are also convenient. Where available, bags having an opaque space for labeling allow you to write more legibly. Before sealing your bags, you can press excess air out of them. This serves two purposes. First, the samples will then take up less space and second, the bags are less likely to puncture or open. Double-bagging samples might be helpful if you are concerned about spills or contamination. Bags should be avoided for samples that are likely to fall apart or become crushed.

Boxes/crates – For large, solid, or rough samples (e.g., rocks), you will need a sturdy container that will not be easily damaged. Keep in mind that boxes will take up more volume than bags and this will influence how many samples you can transport and store.

Coolers – In environmental research, it is often essential to keep samples relatively cool during transport from the field to the laboratory. Coolers (which you may more usually encounter during picnics and tailgates) are frequently relied on. Ice packs (or loose or crushed ice) can be added to lower the temperature further. However, coolers rarely control temperature accurately, rather, they slow the rate of warming the sample might encounter. If temperature is of particular importance, you can keep a thermometer in the cooler, which will help you to monitor conditions. Coolers are really useful, but understand that they are not the same as a refrigerated unit; samples should not be held here for prolonged periods.

Specialist Containers

Gas canisters – Gases extracted from soil or rock can be contained in pressurized cylinders such as Summa or To-Can canisters. Such containers are typically made of stainless steel finished so that the interior is inert using a process known as electropolishing. Steel containers prevent escape of gas and seepage into the container from ambient air. Gas canisters are fitted with secure valves and may also include filters. Be careful not to over-tighten valves as this can damage the seal and risk contamination of your sample. Canisters are evacuated prior to sampling, so no pump is needed in the field. You will also use a manifold to control the rate of gas flow. Typically, when sampling gas from soil pores you will connect the manifold to the sampling port using a Swagelok. You will need both a sample canister and a purge canister. The purge canister will be used to first evacuate gas from the sample tubes. Make sure to clearly label both canisters – you do not want to mix these up. Samples can usually be stored in gas containers for up to 30 d; be sure to consult with your laboratory regarding storage and processing times. Some gas canisters can be reused. They must be fully evacuated under vacuum and then pressurized with nitrogen gas. Heat may also be used, and several repetitions of evacuation–pressurization may be required. Consult the manufacturer's recommendations. Cylinders are usually preferred over gas bags if sampling volatile organic compounds.

Gas bags – Tedlar bags are a popular alternative to gas cylinders. These are made from resilient, inert film, and come fitted with a valve. Bags are easier to transport and often cheaper than cylinders. However, they do not offer the same hold times, and it is recommended to only store samples for up to 3 d in such containers. In addition, a pump is required to extract a sample when using a gas bag as they cannot be pressurized in advance. Keep in mind that they are more vulnerable to puncture during transport than steel cylinders and pack accordingly. Never write directly on a gas sampling bag. Permanent ink can leak through the film and contaminate your sample.

Liquid nitrogen – Some samples require controlled conditions and relatively rapid delivery to the laboratory or to more secure storage to limit the metabolic activity of organisms in the sample. Liquid nitrogen is an inert fluid that maintains an extremely low temperature at atmospheric pressure ($-320\,°F$ or $-196\,°C$). This allows it to be used as a cryogenic fluid that prevents degradation of cells during transport and storage. Commonly, liquid nitrogen is transported in a dewar. These are insulated chambers which may include a pressure release valve (Fig. 4.14). Dewars must always be carried or transported upright. Samples should be

Fig. 4.14 Liquid nitrogen is stored in double-walled containers and maintains samples at $-320\,°F$ or $-196\,°C$. This is crucial for samples which will undergo microbial analysis. *Source:* Sara Vero.

stored in vials or flasks and placed upright within the dewar. Do not use "permanent" markers to label the sample containers – this can come off in contact with liquid nitrogen. When transporting dewars by vehicle, always store it in a separate cab, not in the same space as the driver or passengers. Although it is non-toxic, vapor rapidly expands so if a leak occurs you may be in danger of asphyxiation as oxygen is displaced. This may not be detected until it is too late as liquid nitrogen and its vapor is both odorless and colorless. Under these conditions, a person may pass out or become disoriented and confused. Only work with liquid nitrogen under well ventilated conditions, ideally wearing an oxygen monitor with an alarm setting. Do not work in conditions where the oxygen content is below 19.5% without a breathing apparatus. Insulated gloves, a face shield, and apron must always be worn when handling liquid nitrogen. Upon contact, skin will become burned or frostbitten. These injuries should not be underestimated, and you should seek medical assistance immediately if they occur. Only decant liquid nitrogen slowly as rapid pouring can cause splashing or sudden clouds of vapor. Speak to your lab manager prior to using liquid nitrogen and make sure that you receive full training and appropriate PPE. As with all fieldwork, you must conduct a risk assessment that addresses these hazards.

Soil rings and cores – Soil samples are often taken "intact." This means that a metal ring (often 100 cc volume) is driven into the soil and dug out (Fig. 4.15). In this case, the sampling device (the ring) is in effect, the container for the sample. Removing the ring can result in the sample falling apart. Caps can be placed on the top and bottom of the ring to keep the sample intact during transport and storage.

Fig. 4.15 Bulk density sampling equipment including sampling rings (Ø 5 cm), rubber mallet, trimming knife and wooden block. *Source:* Sara Vero.

Labeling

It is shockingly common for samples to be mislabeled in the field, which can force researchers to ultimately discard them and lose valuable and hard-earned results. This is disheartening and a waste of your time and financial resources. You should label samples in the field at the earliest opportunity. Use permanent, waterproof pens to avoid smudging and running. Sharpies are popular but bear in mind that they may not be completely permanent on all surfaces. It is a good idea to test your pens and labels before going to the field. *Always* label the sample container – not the lid. Someone I once knew labeled over a dozen flasks containing groundwater samples by marking the lids and not the containers. When he prepared to filter them on returning to the laboratory, the lids became mixed up and he had to abandon that round of sampling. There was no alternative way to identify which sample belonged to which location.

You can save significant time in the field by labeling your containers, bags, and boxes beforehand during your preparation at the lab or office. This will also improve clarity, as you can take time to write clearly and legibly. Even better, you may be able to print waterproof stickers or labels.

How should you label your samples? Some laboratories have a management system (such as LIMS – Laboratory Information Management System) which assigns a number or code to samples which allows them to be traced throughout handling and analysis. If you know exactly where and how many samples you intend to collect it may be possible to generate these codes in advance and use in the field. However, it is easy to misidentify or lose track of which samples belong to each location or sampling time. A better approach is to have some logical descriptor that you record in both your field sheets or notebook and on the sample container. Let us consider what should go into these identifiers.

Date/time – This is key, particularly if you are conducting repeated sampling. It allows you to relate the results to the weather conditions or activities at the site at that time, which aids in interpretation. Just as importantly, it facilitates processing samples through laboratory analysis in an appropriate timescale. Certain analyses (e.g., microbial culturing) must be conducted within a limited time from the collection of the sample. This could be because the sample will either change or degrade over time. Consider that laboratories may have hundreds of samples incoming and scheduled for processing each day. Labeling your samples with the date of collection will prevent your samples from getting lost or delayed.

Site – You should identify where your samples were obtained from. For survey-type studies a geographic identifier such as latitude and longitude or GPS co-ordinates may be appropriate. Alternatively, if you have a block or field trial the name of the site might be appropriate. In these cases, you don't have to use the entire name, a letter or numerical identifier will suffice.

Block/plot/rep – Again, in a block or plot study you should use a numerical or letter identifier.

Sampler – If there is more than one individual taking measurements it is helpful to record who took the sample. This allows you to investigate the circumstances or characteristics of the sample, method, location, etc., to aid in your analysis.

That is quite a lot of information to put on a label or bag, but much of this can be abbreviated. Here is an example for herbage from a block experiment. The site name is Tully, the block is number 4, plot 2, and it is the third replicate. So the code is "T_4.2.3" and the date and the sampler's initials are marked below. Other individuals such as laboratory staff or people using the dataset may need to interpret your code. Make sure to record how your sample code can be interpreted and make this information available when needed.

Some laboratories use systems in which sample information is input to a database and unique barcode identifiers are generated. This can be done either before or after fieldwork. Pre-labeling can save you time in the field. When using this approach, make sure you collate all the metadata associated with each sample and barcode. Barcodes are often printed on stickers so check whether these are water-resistant before using outside as some stickers are only suitable for indoor sample handling.

Wildlife

Even if you do not always see them, there are wild animals almost everywhere. For some of you, these animals may be the very subject of your field research and you might actively seek them out. For others, wildlife may be largely irrelevant to your topic of study. Whichever camp you are in, you should consider wild animals, birds, and fish as they pertain to your fieldwork in three ways:

1) **Risk to you/your team** – Although attacks by wild animals in most regions are relatively rare, you may need to be aware of this depending on your location (Fig. 4.16). Remember that animals do not understand what you are doing and may consider you to present a threat, even if you have no interest in them whatsoever. Do not attempt to approach, pet, or otherwise interfere with wild animals (obviously, if you are a wildlife researcher this may not apply). Be aware not only of large animals (bear, buffalo, large cats, etc.) but also snakes and biting or stinging insects. Consult with someone who is familiar with the area beforehand. Animals may also be a potential source or vector for zoonoses.

2) **Risk to the animal or habitat** – Like a good hiker, you should endeavor to do no harm to the environment in which you are conducting your research and particularly, to the creatures which make their home there. Of course, some of your activities will alter the site (digging a pit, installing equipment, etc.); however, you should never destroy a nest, burrow, or den. Remember, many animals and their habitats are protected under law. For example, under the Migratory Bird Treaty Act (1918) prohibits interference with certain species and their nests. Many countries keep detailed online databases of protected habitat areas. You should check whether your intended field-sites fall within these areas and consider how your sites might be affected. Not all fieldwork will be detrimental to wildlife, particularly if conducted in a responsible manner, but if you think that you might disrupt breeding, hunting, or other elements of an animal's lifecycle you should plan (i) an alternative approach or (ii) how to minimize that impact. Never deliberately or inadvertently disrupt or interfere with any animal which is protected under law. You may seek derogation or license for specific activities under your national wildlife legislation and should contact the relevant agencies (often wildlife rangers or departments) as required.

Fig. 4.16 The wildlife you may encounter on fieldwork varies depending on your location. Consider this when planning your trip. *Source:* Jesse Nippert.

Habituation or familiarization occurs when a wild animal or bird becomes accustomed to the presence of humans, and no longer responds with fear or evasion. This is common, particularly when people make food available either by deliberate feeding or by leaving waste or leftovers exposed. While being approached by a wild animal might *seem* charming and make for exciting photos, it is actually extremely harmful. Bears, coyotes, foxes, and large cats have been reported to increasingly venture into human dwellings in rural and suburban areas. Ultimately, this often leads to these animals being killed as they have learned to interact with humans in a disruptive or unsafe manner. Do not contribute to familiarization by leaving food or waste behind you. If you are camping or staying at a site for a prolonged period, store your food supplies in a scentproof container. *Never* deliberately feed any wild animal – you are not "making friends." You are putting yourself and the animal at risk. Thankfully, attacks are relatively rare, but Forrester et al. (2018) estimated that unwanted animal encounters account for 201 deaths annually in the United States and costs US $2 billion in healthcare costs (including one million emergency room visits). Do not become a part of these statistics.

3) **Interference with your site or equipment** – From personal experience and anecdotal reports, this is a common interaction with wildlife that occurs during fieldwork and can be extremely problematic. Rodents, for example, have been known to gnaw wiring of sensors and to burrow around soil monitoring equipment. Their presence also presents a risk of zoonoses as their urine can transmit leptospirosis and other diseases. Birds and rabbits can interfere with vegetation experiments by preferentially eating certain species (e.g. clover in grassland mixed-species swards). Deer and domestic animals such as cattle and sheep can trample plots, pull up pore-water tubes, and knock over temporary or lightweight fencing. These are just a few examples and if you ask around your research institute you are sure to hear plenty more. Keep in mind these risks. You may need to invest in sturdy fencing, netting, or a pest control or wildlife management service. Damage done by animals to equipment can and does occur, and although it is not done with malice, it can be extremely costly or difficult to remedy. Prevention is preferable to a cure, and you should consider potential animal interference during the design of your experiment. Cables, wires, and tubes can be protected from rodent and rabbit gnawing by wrapping in chicken wire or covering with PVC pipe. Plots can be covered with bird netting, although be aware that nets must completely cover the plot(s), should be of sufficiently small gauge to prevent an animal squeezing through, and must be stretched taut over a frame. Poor and ineffective attempts at animal proofing will be easily overcome and are a complete waste of money and effort. For protecting vegetation from bird attack, visual scarers (such as balloons or decoys) and canons or bangers are rarely successful unless accompanied by other methods. Be cautious of over-the-counter pest-control equipment and DIY approaches. There is much to know about the behavior and biology of pest species, and I have come across many unsuccessful attempts to control damage they cause. You may need to consult a wildlife or pest control professional.

A Note on Insects

Working outdoors in almost any environment will bring you into proximity of a wide variety of insects and the hazards and irritations that they pose. Populations are likely to be particularly high during warm and humid weather. Although it is beyond the scope of this manual to give a comprehensive discussion of the wide variety of biting, stinging, and destructive insects, following is a brief discussion of a few issues and precautions.

Biting/stinging insects – Species of biting insects (e.g., fleas, midges, mosquitos, ticks, and spiders) and stinging insects (e.g., wasps, bees, and scorpions) exist in most countries, although in greatly varying numbers and distribution. In many cases, bites and stings are simply irritating, although they can cause substantial discomfort; also, different people exhibit greater or lesser histamine response. However, in some instances, insect attacks can be severe and may cause lasting damage and even fatality. Forrester et al. (2018) indicated that 29.7% of animal-related fatalities in the United States (2008–2015) occurred due to Hymenoptera (bees, wasps, and hornets), a rate that has remained relatively constant over the prior 20 yr. Prior to that, period fatalities had been increasing. What has been responsible for this stabilization? The increased availability of anti-anaphylactic medication such as epinephrine auto-injectors. If you are working in areas where these species are a hazard, you should consider including this in your first aid kit and familiarizing yourself and your teammates with how to use the medication. If only one person understands how to administer the injection, and they themselves are incapacitated, the appropriate treatment may be delayed.

Always be aware of potentially dangerous insects at your field site and carry insect repellent and sting-relieving antihistamine cream in your field kit. If you have an allergy to insect bites or stings make your team aware, carry the appropriate medication, and have an emergency plan. You may not notice insect or arachnid attacks as they occur; for example, tick bites are often only discovered after the fact. If you are operating in an area known to have ticks (grasslands, heather, areas with deer or sheep) you should (i) check your body completely after fieldwork, (ii) carry a tweezers, and (iii) know how to remove a tick. This is done by grasping the head of the tick using a fine-pointed tweezers and lifting it directly away from your skin. Do not grasp the thorax as the mouthparts can be left behind in your skin. Mosquito bites are also of major concern as they are well established vectors of malaria, Zika, West Nile disease, and encephalitis (among other diseases) and their widespread distribution. They may be particularly abundant in warm, humid climates, and near standing water. Insect repellent sprays should be used if working in mosquito habitats and may need to be reapplied throughout the day. Long-sleeved shirts and full-length trousers are recommended, and face nets may be necessary (OSHA-NIOSH, 2016).

Always disinfect and cover bites and stings and seek medical attention if symptoms such as swelling or infection occur.

Crop pests – If you are conducting a crop or vegetation study certain insects can severely harm your research by infesting, consuming, boring, or otherwise damaging your study species. Of course, if your research is to evaluate the efficacy of certain pesticide formulations, then the presence of these insects may not be a bad thing, as it allows you to rigorously examine your treatment. However, in other instances, the arrival of a pest species can confound your work by adding an uncontrolled variable. You may need to abandon plots that are preferentially infested or invest in insect treatments. You should record insect activity in detail (dates, damage done, correct identification of species) and any actions taken in response.

Domestic Animals

It is easy to recognize that wild animals may pose a hazard; however, domestic animals are responsible for a greater proportion of animal-related deaths and injuries (Langley and Morrow, 1997; Forrester et al., 2018). This may be due to our greater contact with pets and livestock but also to lower perceived risk from these familiar species. Of these, dogs are the leading cause, followed by large farm animals such as cows and horses. The farm is in fact the most common site of a mammal-related fatality (35.5%), followed by the home (29.5%) (Langley and Morrow, 1997). For the field researcher, this means that livestock should be carefully considered as potential hazards both if you are directly handling them as part of your experimental design or if they are present at your field sites. They may be harmful through aggressive behavior such as biting, but particularly with larger animals, they can inadvertently crush, knock over, or tread on a person (Fig. 4.17).

Take notice of the location of livestock in the area that you are working in, particularly if they are loose or roaming (e.g., a herd of cattle in a paddock). Many domestic animals are accustomed to handling and may be curious about your activity or assume that you are there to feed or otherwise interact with them. While they may not intend any harm, conducting sampling can be rather more difficult under the close attention of some interested Holsteins. Be aware if animals become too close for you to operate comfortably, or if they become aggressive or intrusive. Behaviors such as snorting, bucking, crowding, or stampeding are obvious signs of danger and you should either move away or if possible, establish a perimeter such as a temporary fence between you and the livestock. Of course, your research might necessitate direct interaction with livestock. If so, you should obtain livestock handling training before beginning your research and take account of them in your hazard and risk assessment. Remember, this is not only for the safety of your team but is also vital for the health and welfare of the animals.

Fig. 4.17 Domestic animals are responsible for a greater proportion of animal-related deaths and injuries than wildlife (Forrester et al., 2018). You also can pose a hazard to farm biosecurity. *Source:* Sara Vero.

Zoonoses and Biosecurity

Zoonoses (singular: zoonosis) are diseases or infections that can be transmitted between humans and animals, including wildlife, pest species, insects, and livestock (Fig. 4.18). Transmission can be animal to animal, human to human, animal to human, or human to animal. It is important to understand that not all zoonoses have an effect on humans, but we can still be carriers (vectors) that can transmit the pathogen between locations. Zoonoses may have relatively minor effects

Fig. 4.18 This gray rat snake is not venomous but be aware of potentially dangerous animals in your area. *Source:* Krista Keels.

such as mild discomfort or even a total lack of discernible symptoms (asymptomatic). Alternatively, they can have severe implications for human and animal health, ecosystem functioning, national or international trade, and food security! For example, the 2001 Hoof and Mouth disease outbreak devastated UK and Irish cattle and sheep herds and had an impact of €169 million on the Irish economy alone. Of course, that is somewhat a worst-case scenario, but recognize that infections and outbreaks may have significant impacts that cannot be easily predicted or controlled.

Broadly, there are three types of zoonotic pathogens you might encounter:

Bacteria – These are single-celled organisms that are ubiquitous to most environments. Although most species are harmless, serious bacterial zoonoses include leptospirosis, campylobacter, and *Escherichia coli*. While bacterial infections can be treated with antibiotics, resistance is developing in some species. It is estimated that more than 60% of bacterial diseases affecting humans are zoonotic (Taylor et al., 2001).

Viruses – These DNA or RNA species can only reproduce within a host cell of another species. They cannot be treated by antibiotics and can cause chronic (prolonged or recurring) illness. Viral zoonoses include bluetongue (a cattle disease), influenzas, West Nile disease, and hemorrhagic fevers (a strain of which is currently causing extensive damage to European rabbit populations).

Fungi – These are a diverse category of organisms, ranging from yeasts to mushrooms to algae. Fungi may form single-celled reproductive spores and remain dormant within the environment until conditions are optimal. Diseases caused by fungi may affect any tissue, but skin and the respiratory system may be particularly at risk. Potential illnesses include various forms of dermatitis and aspergillosis. It should be noted that birds' nests and guano may be associated with particularly high concentrations of fungal spores.

Many zoonoses and their vectors are parasites. Parasites depend on a host to survive (e.g., as a food source for ticks or mosquitos) or reproduce (e.g., protozoa such as toxoplasmosis). While some parasites cause illness themselves, others transmit disease such as the malaria bacteria spread by mosquitos or the α-gal meat allergy acquired from injection of a reaction-triggering carbohydrate by the Lone-star tick.

Transmission of any zoonosis requires contact with the disease-causing agent. Contact can be direct, such as being bitten by a mosquito or handling infected livestock. Indirect contact is perhaps even more common, in which a person is exposed to urine, feces, feathers, or hair, etc. A good example of this is transmission of Weil's disease, a form of leptospirosis carried by rats. Rats often live and urinate in watercourses. This means that many rivers and stream banks will have pathogenic bacteria and/or bacterial spores. For this reason, kayakers and anglers have long been aware of the risks of contracting this disease, and it occurs prevalently in people participating in water sports or who are subject to occupational exposure (Waitkins, 1986; Monahan et al., 2009). They may never even see the rat, but become indirectly exposed to bacteria that was spread by that species. A researcher working in these environments is similarly exposed. It may not be possible to wholly avoid this; if you are a stream researcher, you will probably need to visit a stream. This is known as "occupational exposure." Following the principles of risk assessment, you can minimize the threats posed to you and your team by identifying the potential hazard, evaluating the risks it poses, and implementing a strategy to reduce those risks. These strategies form *biosecurity*. It is important to recognize that secondary exposure is a form of indirect contact. Imagine that you have been exposed to a pathogen; *Toxoplasmosis gondii*, while taking soil cores (*T. gondii* can form resilient spores which may live in soil for months after being shed by a host organism). In this scenario, you are strong and have a healthy, uncompromised immune system, and you do not exhibit any symptoms. However, do you have a partner or a child? Will you shop and dine in public areas? Are you going to visit an elderly relative or attend a conference with thousands of others? You have no way of knowing if you are a potential vector for this disease, or whether these other individuals might be at risk due to their own immune status. *Never* take biosecurity for granted. It puts yourself and others at risk.

Biosecurity refers to the steps taken to prevent transmission of infectious disease or organisms (including zoonoses) and also to limit the spread and establishment of invasive plant and animal species. Biosecurity contains two components: prevention and containment. Prevention involves stopping the exposure of an individual or site to a potentially infectious agent. This implicitly limits the exposure of secondary victims. Secondary exposure occurs when an individual who was not themselves exposed to a pathogen comes in contact with someone who has. Containment is the restriction of the spread of an infection or disease once it has been established. This means that contact with potential secondary victims must be eliminated.

If there are specific pathogens or vectors in your environment (e.g., the Lone-star tick on tall-grass prairie) your research institute should provide safety protocols. If no one in your institute has encountered that hazard before, or you are establishing a new site, you should investigate the correct procedures. Make sure that you and your entire team are familiar with both the hazard and the steps they must take to ensure their safety and that of others. However, there are some basic principles that are applicable in most scenarios.

○ Avoid or minimize contact with potential vectors. Only handle livestock or wild animals if it is an essential part of your experiment. Remember that dead animals may still carry parasites and bacteria and remain a vector.

○ Use appropriate PPE. This will almost always include gloves, and even double-gloving where contamination poses a high risk. Gloves should always be discarded after use and changed between sites. Never open or drive your vehicle while wearing gloves that you have used in the field. Any bacteria or contaminant which may be on the exterior of the gloves are now on your door-handle or steering wheel! This is a common mistake, so institute a best-practice routine. Consistency is key to success.

o Cover all cuts and abrasions immediately, no matter how minor. Again, this is easy to neglect, particularly as minor nicks are common when working outdoors. Open wounds, however, allow pathogens to more easily enter the body. Avoid touching your eyes, nose, or mouth while working. These soft tissues are vulnerable to infection and may allow a pathogen to enter the blood or digestive system. Keep your fingernails trimmed so that dirt cannot lodge underneath.

o If you are working in an area which might have biting insects such mosquitos, or arachnids (including ticks) check for and remove. You may need a teammate to assist (see the buddy-system). Wear long sleeves and trousers when operating in habitats for these species, or if handling animals that carry them such as sheep or deer.

o Wash your hands whenever an opportunity presents; at a rest stop, field station, etc. Steps for proper handwashing are as follows:
1) Remove all jewelry
2) Wet hands thoroughly with running water
3) Apply liquid soap to one cupped palm
4) Wash palm-to-palm, back of hands, interlaced fingers, backs of fingers, both thumbs and wrists
5) Rinse hands under running water and dry thoroughly

o Use a skin disinfectant regularly throughout the day. Consider putting a reminder on your vehicle dashboard or on the front of your notebook. Remember that alcohol-based hand sanitizers are not wholly effective against all pathogens.

o Avoid touching your face, especially your eyes, nose or mouth until you have degloved and disinfected.

o Disinfect tools or equipment regularly, particularly if you are moving between sites.

o If operating on farms or around livestock, always use boot washing facilities wherever they are available. It is good practice to carry a spray bottle of suitable disinfectant in your vehicle to spray your boots before entering or leaving a site. Alternatively, you may use disposable overshoes.

o Monitor the health of yourself, your team, and your family. Consult your doctor if you are concerned.

Harm to health as a result of zoonoses are typically rare as field researchers generally use good sense and implement safety protocols. However, you should never take these risks for granted.

Invasive species are plants and animals that are not native to a location and frequently present significant risks to the health of those ecosystems they invade and to native species. They span a wide range of taxa; from vertebrates (such as the gray squirrel which greatly displaced the red squirrel in Britain and Ireland due to its indiscriminate feeding habits and tolerance for squirrel-pox (Feigna, 2012)), insects such as the Pharaoh ant and both the Chinese and German cockroach (now widespread throughout the United States (Vargo et al., 2014)), fungi such as the crayfish plague (which is lethal to white-clawed crayfish (Oidtmann et al., 2002)) and plants such as giant hogweed or Japanese knotweed (Manchester and Bullock, 2000). In many nations, it is illegal to deliberately or inadvertently contribute to the spread of invasive species. You should check all equipment and vehicles when moving between sites, particularly if a potentially invasive species has been identified. Keeping to the same access path can minimize disturbance to your site and habitats and limit the spread of invasive species.

Let us discuss two examples: one which poses a significant hazard to the field researcher and one which can be inadvertently spread causing threats to the environment. Many other such species exist, and you must consider those in your own region.

Giant hogweed (*Heracleum mantegazzianum* Sommier & Levier) is a large, flowering plant that closely resembles cow parsley. It can reach up to 5 m in height and has a thick, ridged, hollow stem which is green and red with prominent white hairs. Leaves are broad and serrated, and the plant produces an umbel of white flowers in summer. Seeds are spread by wind, water, and inadvertently by humans. This plant is considered to be a noxious weed as its sap is highly phototoxic. Exposure to the sap causes skin to become rapidly photosensitive. A distinctive rash will develop that resembles large blisters or burns. It can be extremely painful and itchy and can recur even years after initial exposure when the afflicted area is exposed to sunlight. Contact with this plant should be rigorously avoided and all steps to limit its spread should be taken.

Japanese knotweed (*Reynoutria japonica* Houtt.) is a bamboo-like, perennial shrub that grows up to 3 m in height. It produces dense stems, an alternating leaf pattern, red buds, and clusters of white flowers. This species produces extensive rhizomes below the surface that allows it to spread extensively throughout an area and outcompete other plants and has been identified as one of the highest risk invasive species. The spread of these rhizomes can measure several meters in diameter, and greater reach may occur where preferential pathways such as pipes or drainage channels are available. It has spread notably along watercourses and roadways. Japanese knotweed damages hard structures, pavements, buildings, and bridges by gouging mortar and brick, lowering the stability of the structure. Even small fragments of rhizome, when transported to a second location can generate new growth, and this has allowed explosive spread of knotweed throughout Europe and the United States. Cutting of rhizomes can actually facilitate the spread of fragments, rather than killing the plant as might be intended.

References

Brooks, A. and Lack, L. (2006). A brief afternoon nap following nocturnal sleep restriction: which nap duration is most recuperative? *Sleep* 29(6), 831–840. doi:https://doi.org/10.1093/sleep/29.6.831

Cooper, J. and Wright, J. (2014). Implementing a buddy system in the workplace. *PMI Global Congress 2014, North America, Phoenix, AZ*. Newtown Square, PA: Project Management Institute.

Dembe, A.E. (2001). The social consequences of occupational injuries and illnesses. *American Journal of Industrial Medicine* 40, 403–417. doi:https://doi.org/10.1002/ajim.1113

Environmental Protection Agency. (2007). *Guidance for preparing Standard Operating Procedures (SOPs)*. EPA QA/G-6. Washington, D.C.: United States EPA Quality Staff.

Feigna, C. (2012). *A study of squirrelpox virus in red and grey squirrels and an investigation of possible routes of transmission*. PhD Thesis. Edinburgh, Scotland: College of Medicine and Veterinary Science. University of Edinburgh.

Forrester, J.A., Weiser, T.G. and Forrester, J.D. (2018). An update on fatalities due to venomous and nonvenomous animals in the United States (2008–2015). *Wilderness & Environmental Medicine* 29, 36–44. doi:10.1016/j.wem.2017.10.004 [erratum: 28(2):284]

Gladwell, M. (2005). *Blink: The Power of Thinking Without Thinking*. New York: Back Bay Books.

Langley, R.L. and Morrow, W.E. (1997). Deaths resulting from animal attacks in the United States. *Wilderness & Environmental Medicine* 8, 8–16. doi:https://doi.org/10.1580/1080-6032(1997)008[0008:DRFAAI]2.3.CO;2

Lima, M.R.M. (2018). Professional online presence and social media. *CSA News* 63(9), 28–29. doi:https://doi.org/10.2134/csa2018.63.0918

Manchester, S.J. and Bullock, J.M. (2000). The impacts of non-native species on UK biodiversity and the effectiveness of control. *Journal of Applied Ecology* 37, 845–864. doi:https://doi.org/10.1046/j.1365-2664.2000.00538.x

Miller, V.G. and Bates, G.P. (2010). Hydration, hydration, hydration. *The Annals of Occupational Hygiene* 54(2), 134–136.

Miller, V.G. and Bates, G.P. (2007). Hydration of outdoor workers in northwest Australia. *Journal of Occupational Health and Safety of Australia and New Zealand* 23, 79–87.

Monahan, A.M., Miller, I.S. and Nally, J.E. (2009). Leptospirosis: risks during recreational activities. *Journal of Applied Microbiology* 107(3), 707–716.

Muzet, A., Nicolas, A., Tassi, P., Dewasmes, G. and Bonneau, A. (1995). Implementation of napping in industry and the problem of sleep inertia. *Journal of Sleep Research* 4(2), 67–69. doi:https://doi.org/10.1111/j.1365-2869.1995.tb00230.x

Oidtmann, B., Heitz, E., Roberts, D. and Hoffman, R.W. (2002). Transmission of crayfish plague. *Diseases of Aquatic Organisms* 52, 159–167. doi:https://doi.org/10.3354/dao052159

PEW Research Center. (2019). *Mobile Technology and Home Broadband*. Washington, DC: PEW Research Center.

REI. (2018). *How Much Should Your Pack Weigh?* Kent, WA: Recreational Equipment, Inc.. https://www.rei.com/blog/camp/how-much-should-your-pack-weigh (Accessed 28 December 2018)

Taylor, L.H., Latham, S.M. and Woolhouse, M.E.J. (2001). Risk factors for human disease emergence. *Philosophical Transactions of the Royal Society, B: Biological Sciences* 356(1411), 983–989.

Transportation Research Board and National Research Council. (2011). *The Effects of Commuting on Pilot Fatigue*. Washington, DC: The National Academies Press. doi:10.17226/13201

USGS. (2006). Section: A national field manual for collection of water-quality data. In *Techniques of Water-Resources Investigations of the U.S. Geological Survey*. Washington, DC: U.S. Geological Survey.

Vargo, E.L., Crissman, J.R., Booth, W., Santangelo, R.G., Mukha, D.V. and Schal, C. (2014). Hierarchical genetic analysis of German Cockroach (Blatta germanica) populations from within buildings to across continents. *PLoS One* 9(7), e102321. doi:https://doi.org/10.1371/journal.pone.0102321

Waitkins, S.A. (1986). Leptospirosis as an occupational disease. *British Journal of Industrial Medicine* 43(11), 721–725.

5

Environmental Monitoring

It is often important to record time-variable parameters such as weather, water discharge, nutrient concentrations, soil data, either because they are the main subject of your research, or to contextualize and understand your observations. While no one rule indicates the appropriate spatial scale or temporal resolution for ecologic or environmental studies, mismatches between data resolution and model requirements can constrain research (Abatzoglou, 2011). You should consider both the spatial and temporal resolution of your data with respect to the intended use, the availability of existing datasets, and the logistical and financial aspects to collecting new information.

At the lower end of the resolution scale, typical, average, or trend data is often provided by local or national meteorological agencies. This information helps the reader of a scientific paper or report because it allows them to make comparisons. If agencies know the general climate of your sites, they can evaluate how your research and findings might translate into their area and to consider what differences might exist. Data derived from individual weather stations may not adequately represent exact conditions distant from their immediate vicinity. If publicly sourced data is used, it may require processing to make it suitable for your target application (Abatzoglou, 2011) such as development of gridded datasets that interpolate data from multiple stations.

At high-resolution, daily or sub-daily data recorded at your specific site might be crucial for understanding spatiotemporally variable behavior. This may require installation of monitoring equipment such as weather stations, soil monitoring arrays, autosamplers, or in-stream equipment.

You should think carefully about your data requirements. Some pros and cons of high- and low-resolution data are described in Table 5.1; however, it may be helpful for you to prepare your own list when designing your project.

Resolution, Precision, Accuracy, and Hysteresis

Resolution, precision, and accuracy are often incorrectly used interchangeably when dealing with environmental data. They actually refer to different but related behaviors. It is important to understand these terms when interpreting sensor data.

Resolution is how fine or detailed a measurement is delivered by a device. For example, a balance that measures to four decimal places (e.g., 1.0298) has a greater resolution than one which measures to only two (e.g., 1.03). Depending on the parameter and the intended use of the data, fine resolution may be more or less important, but should always be noted.

Fieldwork Ready: An Introductory Guide to Field Research for Agriculture, Environment, and Soil Scientists,
First Edition. Sara E. Vero.
© 2021 American Society of Agronomy, Inc., Crop Science Society of America, Inc., and Soil Science Society of America, Inc. Published 2021 by John Wiley & Sons, Inc.
doi:10.2134/fieldwork.c5

Table 5.1 Pros and Cons of low and high data resolution.

Low-resolution data		High-resolution data	
Pro	**Con**	**Pro**	**Con**
Inexpensive or free	Rarely specific to your site	Site-specific	Expensive (cost of equipment)
Readily available (e.g., from national meteorological services)	Coarse temporal and spatial resolution may mask dynamic behaviors (e.g., hydrology)	Temporal resolution may be controlled	Time-consuming/ laborious to install and maintain arrays, and thereafter, to process data
Validated or quality controlled by meteorological agencies or other providers	Nontransparent methods, sources and meta-data	Option to control/ check quality	
		Full control over parameters and locations	

Accuracy is the random error between reality and the measured value as a result of "noise" (unexplained variation). Accuracy is expressed as a deviation from a measured value (e.g., $\pm 1\%$ (relative error) or ± 1 g (absolute error)). When interpreting data, you should understand that the reading reports the measured variable allowing for error within this range. In other words, if your device reported a value of 10 g and the precision is $\pm 1\%$ in actuality the variable is between 9.9 and 10.1 g. Or in the case of absolute error, if your device reported a value of 10 g and a precision of ± 1 g, the variable is actually between 9 and 11 g.

Some recommendations regarding the ranges of accuracy and precision for meteorological monitoring are given in Meteorological Monitoring Guidance for Regulatory Monitoring Applications (U.S. Environmental Protection Agency, 2000), which may have utility in environmental research.

Precision is the ability of a sensor or other device to produce the same result or reading consistently. Note that the value may not be accurate, but it can be precise. In other words, the device might always report values that are tightly grouped but are not close to reality. This can be improved by calibrating the device.

Think of some archers shooting at a target. The first archer is accurate and precise, hitting the bull's eye five times (Fig. 5.1a). The second archer is precise, but not accurate, all their arrows are clustered together 2 in. away from the bull's eye (Fig. 5.1b). The third archer hits the bull's eye with one arrow, but the other shots are scattered; they are not precise, and their accuracy is low (Fig. 5.1c).

Improving these three parameters will allow measurements that better reflect the variable being measured. In most cases, resolution is improved by switching to a more sensitive sensor, although

(a)   (b)   (c)

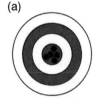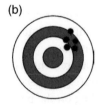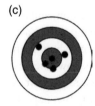

Fig. 5.1 (a) Accurate and precise; all the arrows hit the bull's eye, (b) precise but not accurate; all the arrows are close together but distant from the bull's eye, and (c) neither precise nor accurate; the arrows are neither close to each other nor the bull's eye. *Source:* Sara Vero.

this may be more expensive. You should consider what the application of the required data is and tailor your sensor selection to that. Higher resolution is not always best. For example, if you want to know soil salinity across a field of row crops, is it more important to have sensors that measure to a resolution of $0.001\,dS\,m^{-1}$ or to $0.0001\,dS\,m^{-1}$? Or is more important to have a larger number of sensors with lower resolution so that the spatial heterogeneity of the field can be captured? This is a case-by-case decision.

There are simple ways to improve accuracy and precision:

1) **Optimum installation** – A poorly installed sensor is unlikely to give accurate results. It can become damaged, impaired, or simply measure an area that is not reflective of your target or which is unduly influenced by other variables. Let's consider some examples. A moisture probe is inserted vertically and not in adequate contact with the surrounding soil. Water can then preferentially flow down and along the probes, leading to a higher measured value than is actually reflective of that horizon. If a pyranometer for measuring solar radiation is placed near a lake, light reflected from the water surface will lead to unrepresentative values. Always carefully read the manufacturer's instructions prior to installation and take your time in the field. Time spent on good installation will save time spent troubleshooting later.

2) **Calibration** – Calibration is essential to ensure accuracy as instruments can deviate from factory standards due to damage, age, installation, and drift. Devices are calibrated to within a "tolerance range." This is the range of acceptable variance relative to a known standard. Methods of calibration include:

   a) A standard of a known value can be measured, and the reading of your device adjusted until it conforms. For example, a pH sensor can be used to measure the pH of a chemical standard and adjusted as necessary. Chemical standards must be correctly stored and will typically have an expiration date. Always check that your standard is within that date before beginning a calibration.

   b) The measurement of Device A can be compared against that of Device B which is of known accuracy. For example, a measurement taken in the field under environmental conditions can be examined versus one taken with a laboratory device under controlled conditions.

   c) Several measurements can be taken along a range of values and a curve is fitted. For example, if calibrating a moisture sensor, you might take a measurement in oven dry soil, at moisture contents of 5%, 10%, 15%, and so forth. Using this curve, you can determine the accuracy of the measurement based on how the goodness of fit and also identify the regression equation by which measured variables should be adjusted.

Most equipment will come with detailed steps for calibration and standards such as certified weights or solutions may be available to purchase. It is essential that calibration is conducted under controlled conditions. You must measure the variance of the sensor or device, not that imposed by the transient environment in which they operate. Many modern devices that rely on electronic rather than manual readings will have automatic calibration functions by which the user simply follows a sequence of steps that allow adjustments to be made internally. This removes much user error but perhaps limits your understanding of where or how the device deviates. This may not be problematic but take care to study how exactly these changes are made.

Hysteresis is a non-linear behavior in which physical responses differ depending on the history or direction in which a force, pressure, or other input is exerted on a system. A common example in environmental science is that the volumetric water content of a soil at a specific matric potential

---

When Should You Calibrate?

- When you first purchase a device.
- Prior to deploying an in situ sensor or probe.
- At pre-arranged frequencies. These may be specified by the manufacturer or are standard to your laboratory or research institute. Think of this like servicing your car or seeing your dentist. Don't wait until there is a problem!
- When recommissioning a device after storage.
- After a project or measuring regime ends – this allows you to review the measured results and address any errors that may have been recorded.
- If the device has been dropped, vibrated, or bumped. Some devices are more or less resilient than others, but "minor" impacts can have significant effects on measurements.

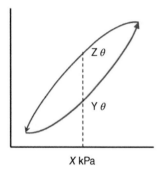

Fig. 5.2 Soil moisture hysteresis means that at a certain matric potential (kPa) the actual moisture content could be either $Y\theta$ or $Z\theta$. *Source:* Sara Vero.

is different depending on whether the soil is either wetting or drying Fig. 5.2. Hysteresis is challenging because it hampers direct inference of a value or property based on another variable, unless you take antecedent conditions into account. Considering the previous example, at X matric potential the soil could be at either Y or Z volumetric moisture content. The significance of these differences will vary depending on the substance and parameter being measured. Antecedent data should always be considered when interpreting potentially hysteretic parameters and prolonged monitoring can help you to understand these cycles in the environment. Stream sediment transport and nutrient dynamics are also hysteretic, and the direction and shape of hysteresis loops have been used to interpret the sources (Sherriff et al., 2016; Mellander et al., 2015). If your parameter of interest is likely to behave with hysteresis, it becomes particularly important to establish baseline measurements prior to your main study period.

Monitoring Arrays: An Overview

There are several different types of monitoring arrays, capable of measuring and recording weather, gas, soil moisture, solutes, or in-stream parameters (Fig 5.3). However, the basic concept is the same; either one or several sensors or devices measuring the target parameters are controlled by a central datalogger (computer) that also records the data. This section will not tell you how to program your array as this varies greatly between manufacturers and specific devices and most suppliers will provide detailed instructions and even downloadable programs which you can edit and adapt. Rather, let us discuss the different components of a monitoring array and give some advice on how to install them. Although many different manufacturers and contractors who can design and install monitoring arrays, it is helpful to have at least a basic understanding of these fixtures as they are a common feature of environmental and agricultural fieldwork.

Data-logger and sensor technology is becoming increasingly user-friendly, with many commercially available requiring minimal (or no) configuration or wiring. However, it is still often

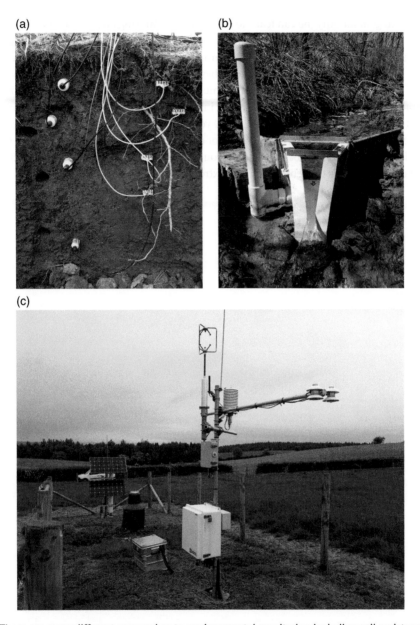

Fig. 5.3 There are many different approaches to environmental monitoring including soil moisture sensors at various depths (a), streamflow monitoring using flumes or weirs (b), or weather stations (c).
Sources: Brandon Forsythe and Sara Vero.

necessary to wire, program, assemble, and install arrays. This allows older equipment to be reused at a new site or enables more complex logger-device configurations than might otherwise be possible. Before preparing your equipment, there are two issues that should be considered. Safety should be foremost; never work on a live circuit, always wear safety goggles, ground your equipment correctly, and keep your workbench tidy and dry. Second, data-logger setup can be challenging, particularly if you are not familiar with the equipment or with basic electronics and programming in general. Anyone can learn to establish a monitoring array, so do not be intimidated if it seems complex at first. Ideally, you should consult with your supervisor or other researchers in your university or research center who have experience with the equipment and can guide or train you. It may be helpful to take a short course in basic electronics or specifically in environmental monitoring. If you have sufficient budget, some equipment manufacturers and independent research training agencies offer courses that can instruct you. This could be a helpful investment, as it is easy to waste considerable time learning by yourself, and these courses are taught by experts. Equipment suppliers typically provide helplines or consultants that you can reach out to for advice on specific issues and who may be able to talk you through the setup process. Finally, discussion boards and online tutorials or webinars can be extremely helpful.

Components

Logger-Box or Enclosure

The logger-box is a rugged container, usually plastic or coated metal that is used to store the data-logger and battery. These boxes can be purchased from logger suppliers or fashioned from suitable materials. Boxes are typically waterproof, and you may include a bagged desiccant such as silica and a humidity indicator to identify water intrusion and offset the risk of electronic equipment becoming damp. Logger-boxes can be affixed to poles, tripods, or walls to keep them off the ground and may also be fitted with locks. It is worthwhile to invest in a sturdy logger-box and to install it properly.

Monitoring of streams and rivers often involves large equipment including pumps, autosamplers, tanks, telemetry, and other devices. These will not fit in a mounted logger-box. Instead, waterproof sheds (commonly referred to as "kiosks") are used. These stations protect the equipment from the elements and provide shelter to the researcher as they work. They also may prevent tampering by the public or interference from animals. Kiosks can be purchased from many equipment supply companies. Kiosks should be installed on a solid, stable base.

Case Study – The Importance of Logger-Boxes

During my PhD research, one of my soil sensor arrays was accidentally struck by a contractor using a tractor-mounter hedge cutter (essentially, a chainsaw) who did not see it as he was working. The logger-box was cracked in several places and one of the mounting brackets was severed, however, the data-logger was unharmed, only one measurement was missed (I was using a 10-min resolution) and none of the attached sensors failed. The damage without such a high-quality logger-box would likely have been catastrophic. It is worth investing in a sturdy enclosure!

The Data-Logger

The data-logger (or simply, logger) is the brain of the monitoring station (Fig. 5.4). It is a processor which may have memory and/or transmission capability which can be used to access your data, either by downloading via direct connection with the logger itself, by uploading to the cloud, or by direct transmission using telemetry or other signaling. Loggers do not take the measurements themselves, but rather, collect and store the data from the connected sensors or devices. Loggers instruct the sensors when or how to take the measurement via instrumentation protocols. Serial Digital Interface at 1200 Baud (SDI-12) is a particularly common communications protocol. This protocol was designed in 1988 by the US Geological Survey's Hydrologic Instrumentation Facility and a group of private firms, many of whom continue to be leaders in environmental monitoring. Essentially, the SDI-12 protocol assigns "addresses" to each sensor in the array. This allows multiple sensors to be connected to and controlled by a single logger. Sensors and loggers produced by many different companies can interact using SDI-12 as a common language, providing variety and flexibility to design arrays specific to your requirements without being constrained to a single equipment manufacturer. A useful website for further information on the SDI-12 protocol is http://www.sdi-12.org/.

While data-logger design varies by make and model, there are some connections which are common (Table 5.2).

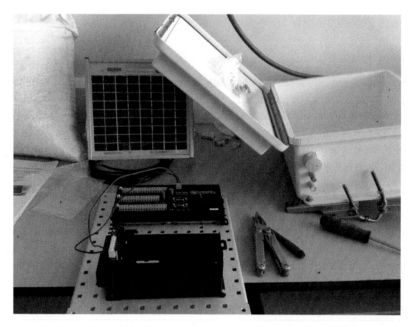

Fig. 5.4 Setting up a data logger requires programming using software such as CR Basic or Edlog, and wiring of peripherals such as the power source and sensors (shown here). Always prepare your logger and test that the program is working prior to installation in the field. *Source:* Sara Vero.

Table 5.2 Summary of common connection types and descriptions.

Connection type	Description
Switched voltage excitation outputs	These connections are used for devices that measure voltage resistance such as volumetric soil moisture sensors. Put simply, the logger triggers a pulse to be sent along the probe and the speed at which it travels or reflects is used to calculate resistance, which in this example, corresponds to water content
Input/output ports	These communicate instructions from the data-logger to the sensors and readings from the sensors back to the logger
Continuous terminals (5 or 12V)	These supply continuous voltage to connected devices/sensors (e.g., radiometers or other devices that incorporate a heating element)
Switched terminals (12V)	These supply power to devices/sensors at programmed intervals during which measurements are taken (e.g., a humidity probe or matric potential sensor)
Ethernet/memory card/ USB ports	These ports enable data storage or connection to a laptop for programming and/or downloading
Multiplexers (MUX)	These are extended terminals that can be connected to the primary data-logger to enable additional devices/sensors to be attached. Multiplexers route multiple peripheral devices through a single port by receiving information from each input line in relay (Fig. 5.5)
Jack-connectors	Although many sensors are wired to their loggers via pigtail connections, some devices are plugged directly using jacks (a simple plug like you would see on earphones). These are easy to connect but may restrict the variety or number of devices you can use. If your sensor comes equipped with a jack and you wish to connect to a nonjack port you can simply snip the connector off with a scissors, strip the wires and route them to the correct ports (Fig. 5.6a). You may tin the exposed wire by applying a small amount of solder prior to connecting to the relevant ports (Fig. 5.6b). This will reduce fraying and allow neater connections

(a) (b)

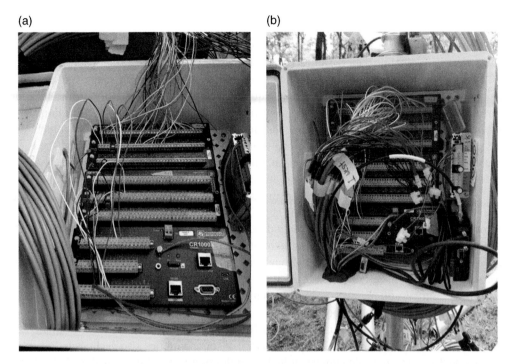

Fig. 5.5 Multiplexers (shown above the datalogger) are used to increase the number of sensors or devices which may communicate with a single datalogger. *Source:* Brandon Forsythe.

(a) (b)

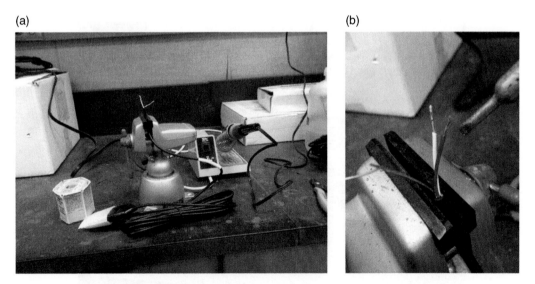

Fig. 5.6 A thin layer of solder can be applied to the exposed ends of wire. This prevents fraying and helps establish a good connection with the datalogger. *Source:* Sara Vero

Power

There are generally three options for powering your data-loggers and monitoring arrays; AC (mains) electricity, alkaline batteries or rechargeable batteries (often with a solar panel).

AC power is relied on for large and demanding arrays such as river monitoring stations or long-term installations. Establishing an AC connection requires permit and registration with the local or national electricity provider, initiation (and payment) of power usage fees, and must be done by a registered or licensed electrician. In other words, this is not an afternoon job for a PhD student. Of course, if the equipment is planned for a long-term or permanent experiment or if the site is on premises belonging to your research institution, this can be a suitable option and may be essential if your equipment has a large power demand. A common example is a refrigerated autosampler.

With battery-powered arrays, you must ensure that new or rechargeable batteries are available and should check power supply regularly (Fig 5.7). If you are manually downloading, be aware that should battery power fail in between downloads you may miss multiple measurements, leading to gaps in your data. If you have telemetry or other remote connection you will be alerted to a break in the data feed when you log on, although you may not know whether it is due to a lack of power or some other issue. You should always have a spare, fully charged battery available to switch if necessary. Caution: batteries are particularly prone to theft and it is advised that you securely lock these in your logger-box.

Alkaline batteries (which you can purchase at hardware stores or even supermarkets) are not rechargeable and can be expensive. The performance of these batteries also declines in low temperatures, so if you are in a cold climate, at high altitudes or during winter you should check the arrays more frequently and be prepared with replacements. Do not mix old and new alkaline batteries as this can cause them to leak caustic battery fluid. European Union member states are com-

Fig. 5.7 Powering your equipment in the field can be challenging. If you are not using solar power you need to check the power supply and replace the batteries regularly. Remember that cold weather may influence stored power. *Source:* Sara Vero.

mitted to the Waste Electrical and Electronic Equipment (WEEE) Directive 2003 (Directive 2012/19/EU). Under this legislation nations must provide recycling facilities to limit hazardous waste electronics in the domestic waste stream. Always dispose of used batteries through a recycling facility and not in general waste. Never leave a used battery on site. These can leach lead and cadmium into the environment (Panero et al., 1995).

Rechargeable batteries are often the preferred option for powering small to medium monitoring arrays, particularly in conjunction with solar panels. Rechargeable batteries have a "cyclic service life" – a number of recharge cycles until it drops to an unacceptably low capacity (Fig. 5.7). As the battery ages its output current (Ampere hour – AHr) will decline. This decline is accelerated by cold temperatures. You can check the estimated service life in the battery documentation. The AHr of your batteries can be checked using an ammeter. You should discard rechargeable batteries once AHr declines to below 60% of its rated capacity (Campbell Scientific, 2011).

Solar panels can typically supply up to 85 W, although more than one panel can be connected to a regulator if there is greater demand (Fig. 5.8). Panels provide a trickle charge to the battery dur-

Fig. 5.8 Solar panels, when correctly installed, can be used to power rechargeable batteries and reduce the need for frequent site visits. *Source:* Sara Vero.

ing daylight hours. The regulator is required to moderate the power input from the solar panel to the battery. Many modern solar panels have built-in regulators, or regulators are incorporated within the data-logger or battery pack. You should consult the user manuals or brochures. The increased efficacy and competitive price of solar panels has undoubtedly contributed to more widespread environmental monitoring and allows remote areas to be instrumented. Large panels are even suitable for demanding arrays such as eddy covariance towers. There are certain details which you should consider when selecting solar equipment. If your array is located in areas with limited light, you should ensure that the battery has sufficient capacity to power the logger and sensors throughout the low-light period, or until your next visit. This is known as "reserve capacity." Your battery must have sufficient capacity to support demand, even when the resupply might be low, for example, on cloudy or dark days. Remember, at different seasons, the recharge of your battery may be curtailed due to short daylight hours.

When designing your array, you must calculate the power requirements so that you can select the appropriate battery (and solar panels). The demand of your array is the total current drain of the data-logger and all connected sensors, multiplexers, and communication devices (Fig. 5.9). The demand of the data-logger is influenced by its activity; in other words, the scan rate (how frequently the logger communicates with the sensors) and level of data processing it is engaged in. Suppliers will indicate the quiescent and active power demands for their loggers, but your programming will dictate where within this range your equipment will operate. This also applies to multiplexers.

Fig. 5.9 Monitoring arrays include sensors, a datalogger and a power source. *Source:* Sara Vero.

Many sensors have very low current drains and can be considered as negligible in your calculations; however, some devices such as thermistors or pressure transducers have a significant and even constant demand. Always check the power requirements of your selected sensors. Communication devices such as phone modems or transceivers only draw power when they are actively transmitting, so naturally will be influenced by the frequency of communication. Note that if you are using telemetry, the data-logger will also be active during this period, in addition to during its scan frequency.

Power Calculation – Example

You want to install a soil monitoring array consisting of a CR1000 data-logger (Campbell Scientific), six 5TE soil moisture sensors (Meter) and a COM220 phone modem (Campbell Scientific). You program the sensors to take a measurement every 15 min, and results will be transmitted at noon and midnight. Let's imagine that each phone transmission lasts 5 min and that the scan rate of the data-logger is every 30 s.

$$\text{Demand} = \frac{(\text{Duration})(\text{Drain})}{\text{Time}}$$

Based on these three components of your array (Table 5.3), the average current drain is:

$$0.66 \text{ mA} + 0.19 \text{ mA} + 0.29 \text{ mA} = 1.14 \text{ mA}$$

Table 5.3 Example power demand for data-logger and volumetric moisture sensors.

	Quiescent		Active		
	Drain (mA)	Duration	Drain (mA)	Duration	Current drain (mA)
CR1000 (scan)	0.6	29.8 s	10	0.2 s	0.66
5TE sensors (×6)	0.03	899.8 s	10	0.2 s	0.19
COM 220	0.012	1430 min	40	10 min	0.29

Source: (Adapted from Campbell Scientific, 2011).

Based on this calculation your array needs 1.14 mA, or 0.00114 A. If your alkaline battery has a capacity of 7 AHr, then it can power your array for up to 6110 h, or 254 d. This may be reduced in cold conditions.

If using a solar panel to charge a battery pack, you must also allow for reserve time. This is surplus power required to fuel the equipment during low light conditions. Suggested reserve times depend on the latitude at which your array is located. For 0°–30° allow 144–168 h, for 30°–50° allow 288–336 h, and for 50°–60° allow 432 h (Campbell Scientific, 2011). In polar regions, 8760 h are required. Let us imagine your array is going to be located at a field site on a dairy farm just north of Des Moines, IA. The latitude is 41.91°. Let us allow the maximum reserve time for safety and allow up to 80% discharge of the battery (0.8).

$$\text{Required Battery Capacity (AHr)} = (\text{demand})(\text{reserve time})/(0.8)$$
$$(0.00114)(168)/(0.8) = 0.2415 \text{AHr}$$

This is well within the capacity of a rechargeable battery with 7 AHr. The AHr required (24 h) is therefore 5.79 AHr d^{-1}. Now, you need to select a complementary solar panel. Let us estimate 6 h daylight per day (of course, this will vary throughout the year).

$$\text{Solar Power Current} > \frac{\left(\text{Total Power Demand} * 1.2\right)}{\text{hrs of daylight}}$$

Your required current is therefore $(5.79 \times 1.2)/6 = 1.15\,\text{A}$. This could be delivered by an SP20 model solar panel which delivers a peak current of 1.19 A. It is worth taking the time to make these calculations prior to purchasing equipment, particularly if a large number of sensors are used, or if individual peripherals such as transceivers have particularly large power requirements. In our example, the logger and sensors were essential, but the phone modem, which may not have been essential, was responsible for just over a quarter of the power demand.

Solar panels should be installed in unshaded areas, angled at 15° from horizontal and facing whichever direction receives most direct sunlight. You may need to adjust this angle to prevent snow accumulation. Panels should be located away from shade and shadows.

It is vital to correctly earth (or "ground") your array to prevent damage to your logger and sensors in the event of lighting strike and to limit electrostatic noise. There will be a copper nut (a "ground lug") on the data-logger to which you should attach an eight-gauge wire. This wire should be routed out of the logger-box to a copper or steel stake (a "grounding rod") and driven firmly into the soil. This will simultaneously earth the peripheral sensors, as each will have an earth wire that is routed via the logger.

Consistency of power supply is important: your careful design, investment, and installation will be wasted if your monitoring array fails to function due to inconsistent power. Environmental monitoring equipment is often intended to reduce the frequency of site visits, or to record over prolonged periods. It is therefore crucial that it does not fail in your absence. You can check the voltage of batteries in the field using a handheld voltmeter (Fig. 5.10) and replace as needed. Data-loggers frequently have a small internal battery which will prevent memory loss in the event of power failure, but these may not wholly protect you and you should download data regularly. These internal batteries may also need to be replaced; check the datalogger specifications. Power surges can also be problematic as spikes in voltage can damage both the logger and peripheral sensors. This can be offset by properly earthing your arrays, or by use of power regulators or surge protectors. These work by buffering the voltage to within a specified range. Surge protection and regulators are incorporated within some loggers and devices but can also be purchased as separate components. Solar panels, for example, often include surge protection to mitigate the variability in power supply.

Fig. 5.10 A battery tester or multi-meter is a useful piece of equipment for checking your power supply in the field. Simply connect to the positive and negative terminals and reading the output on the display. *Source:* Sara Vero.

Sensors

There are a wide range of sensors available commercially for measuring soil, air, and water properties and dynamics. Just as there are different makes and models of cars, there are often several similar devices available from various companies for measuring an individual parameter, but which differ in capacity, accuracy, sensitivity, size, or predicted lifespan. Furthermore, some sensors measure several parameters or may allow secondary parameters to be calculated from the raw data. You should study the specifications when selecting sensors and consult the manufacturer for further information. In most cases, if you can thoroughly describe your experimental design and the environmental conditions in which the sensors will be installed, they will offer advice and support. There is a large community of sensor users and programmers online which provide a wealth of knowledge and advice. Consider reaching out to these individuals when planning your arrays, or if you have questions regarding troubleshooting. Look for designs and specifications of monitoring arrays in the scientific literature. This will help you to determine approaches which are consistent within your specific field.

Although other sensors exist, here are some which you might commonly encounter.

Air, Soil, and Water Temperature

Temperature sensors generally fall into two types, thermistors and thermocouples. A thermistor is a component incorporating a metal that has a known resistance to voltage. As temperature increases, the resistance decreases in a nonlinear fashion (this is why thermistors should be calibrated although suppliers will often indicate the range of the factory calibration). The symbol for thermistors is shown in Fig. 5.11.

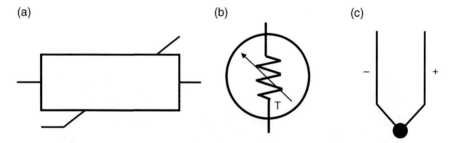

Fig. 5.11 Thermistor (a – European symbol, b – US symbol) and thermocouple (c). These will indicate how a sensor is taking the measurement, although it will also be indicated in the product literature..

Thermocouple sensors consist of two strips of dissimilar metals connected at one end. When one end of the thermocouple is heated, it causes a current to pass through the circuit that can be measured by a voltmeter and related back to temperature. Thermocouple-type sensors are rugged, durable, and capable of measurements over a wide temperature range, although may have lower accuracy than thermistor-type sensors.

When measuring air temperature, it is best practice to house the sensor in a ventilated radiation shield to prevent heating from direct sunlight. Typically, air sensors are mounted between 4 and 6 ft above ground (Pica, 2018) – always record the exact height of your sensor. The mesonet network, for example, stipulates a height of 1.5 m (4.9 ft). Avoid installing sensors near to potential heat islands (such as industrial buildings or livestock housing), shaded areas, open water, or tall vegetation. Regarding soil temperature probes, always record the depth of each sensor. You can install at whatever depth is of interest to you (e.g., in individual soil horizons), but it is common to use depths of either 5 or 10 cm (Petropoulos and McCalmont, 2017; Bell et al., 2013; Brock et al., 1995; and others). Point measurements of stream temperature can be taken across a watercourse using handheld probes which should be submerged and allowed to equilibrate. Changes in temperature along a watercourse can be indicative of anthropogenic inputs such as wastewater discharge points (Kinouchi et al., 2007). Changes (either warming or cooling) can also be indicative of groundwater contribution (Kalbus et al., 2006). Groundwater is typically buffered from sudden temperature changes by the overlying soil and bedrock and remains relatively stable over time, and so may be either cooler or warmer than the surface waterbody it feeds, depending on time of year.

It is also common to include a fixed sensor at stream outlets or monitoring stations. In these cases, the depth of the stream or river will influence the positioning of your sensor or sensors. For relatively shallow streams, a single sensor may suffice, however, in deep rivers where there may be temperature gradients depending on surface warming and subsurface currents, multiple sensors at different depths may be needed. This should be decided on a case-by-case basis, and you should also consider the purpose of your measurement. Sensors can be affixed to stakes or rebar driven into the bed or bank or to a bridge strut, or alternatively can be weighted or tethered.

Stage

The height of water above a fixed datum (stage) can be measured simply using a staff gauge, which is a graduated bar or pole (Fig. 5.12). For continuous recording a pressure transducer connected to a data-logger is required. These measure the weight or pressure of the water column above the

Fig. 5.12 Staff gauges like this allow height of water to be recorded, which allows discharge to be calculated. *Source:* Sara Vero.

sensor, which remains at a fixed depth. It is typical to contain the transducer within a chamber known as a "stilling well." Stream water can pass in and out via open ports so that the water level is equal to that outside of the chamber. This allows accurate measurements to be taken by the transducer (or any other sensor) without influence from turbidity or wave action.

Humidity

Humidity is the amount of water vapor present in the air and is a critical parameter influencing evapotranspiration rates and initiation of precipitation. Humidity is typically measured using a capacitance-type sensor. These sensors consist of two conductors separated by a nonconductive membrane. Depending on the air humidity, moisture will accumulate on the membrane, thereby influencing the voltage between the conductors. Most commercial humidity probes incorporate a temperature sensor. Humidity (and temperature) sensors should be enclosed within a radiation shield to limit thermal radiation (U.S. Environmental Protection Agency, 2000).

Wind Speed

Wind speed is measured using an anemometer (Fig. 5.13). There are several versions of this device, from relatively simple apparatus which log the number of rotations a vane, blade, or cup makes in a given time to pressure or sonic devices. Regardless of device type, location is crucial as wind speed will vary with height, proximity of obstructions such as buildings or trees, and will also be influenced by turbulence. It is common to position anemometers at 10 m above ground level, although greater heights may be needed if hard obstructions are nearby (within 50 m) (U.S. Environmental Protection Agency, 2000).

Fig. 5.13 An anemometer is used to measure wind speed. *Source:* Jesse Nippert.

Rain Gauges

A variety of rain gauge designs exist, including simple graduated collectors, weighing devices, and tipping buckets (Figs. 5.14 and 5.15). Graduated collectors funnel rainfall into a cylinder of known

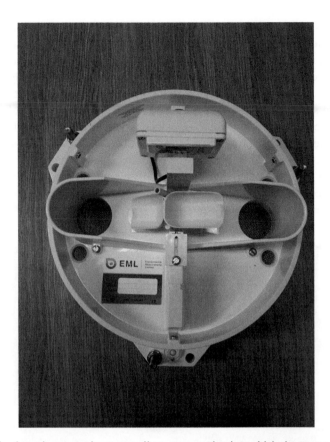

Fig. 5.14 Tipping bucket rain gauges have a small see-saw mechanism which tips over when each bucket is full. The number of tips is recorded by a datalogger. It is essential that the gauge is level and stable. *Source:* Sara Vero.

Fig. 5.15 Preparing a rain gauge for deployment. *Source:* Sara Vero.

volume. These devices are not self-recording and so must be inspected and manually recorded and emptied. Their greatest limitations are the need for regular observation and also, they may overfill subsequent to heavy rainfall or if not tended to frequently enough. A tipping bucket rain gauge can help you to overcome these limitations. These consist of two small and carefully balanced buckets on a fulcrum; like a child's see-saw. When one bucket is full, it tips over, empties, and allows the alternate bucket to start collecting water. Each "tip" is recorded by the data-logger, which can thereby calculate the quantity of rainfall. Rain gauges should be positioned at least four times the distance from the height of any obstruction. They must be installed on a level base. Use a spirit level to check this at the time of installation.

When purchasing precipitation monitoring or collection equipment, check with the manufacturer whether they conform to National Weather Service (or local meteorological agency) specifications (Fig. 5.16).

Fig. 5.16 Snow can interfere with rain gauges. Take note if/when this happens, record the height or weight of accumulated snow and remove. If you are in an area in which this is common, you might consider also installing a snow gauge. *Source:* Brandon Forsythe.

Volumetric Soil Water Content

Volumetric water content (VWC) (θ) is the volume of water relative to the total volume of the soil (as opposed to gravimetric water content, which is its mass). Soil consists of solid particles interspersed with pores of various sizes and complexities. These pores are at any given time filled with both water and gas. It is assumed that (unless compacted or otherwise disrupted) the solid proportion remains fixed, but the contents of the pores will fluctuate over time depending on precipitation and drainage. Remote sensing is becoming increasingly popular as a means of assessing moisture held in the soil; however, there remain limitations around cost, spatial and temporal resolution (including depth resolution), and availability. For reliable and site-specific measurements, sensors are still preferred. Several companies produce accurate soil moisture sensors suitable for pot, plot, and profile installation, along with some novel installation tools. Many of these sensors work via time domain reflectometry. This technique measures the dielectric permittivity of the surrounding soil, that is, the amount of charge it can hold by sending an electromagnetic pulse along two prongs. This is related to the amount of pore space filled with water rather than gas (although temperature and salinity can also exert an influence). Some devices have a third prong which contains a temperature sensor which increases the accuracy of the water content measurements. Other approaches include neutron probes which scatter and detect α particles emitted from radioactive pellets.

Soil moisture sensors should be installed reasonably in advance of the intended monitoring period. This is for two main reasons. First, it allows the sensor to become embedded more naturally within the soil. It is not unusual to see fluctuations immediately after installation as the soil settles around the sensor prongs. This is particularly common if a pit or trench was excavated and borehole installation may be less disruptive to the integrity of the soil profile. Second, advance installation allows you to collect baseline measurements, which are often helpful in characterizing the site and contextualizing your study.

Matric Potential

Matric potential (ψ) is the measure of how tightly water in the soil system is bound to solid particles. This controls the movement of that water, such as via percolation, lateral flow, or evapotranspiration. Matric potential ranges from 0 (saturation) to −1500 kPa (permanent wilting point), with field capacity considered at c. −33 kPa (although field capacity is a rather general approximation). In simple terms, these sensors work by measuring the difference in dielectic permittivity between ceramic disks and the surrounding soil. The ceramic head of the samplers should be soaked in water for 24 h prior to installation. Once they have soaked, cover with tinfoil to ensure that they remain saturated during transit to the field.

Tensiometers are an alternative to sensors. A tensiometer is a water-filled tube (usually glass or plastic) that ends in a porous ceramic cap. This is saturated and inserted with good surface contact into the soil. The other end of the device is equipped with a pressure transducer and logger or simply a manometer. The relative wetness or dryness of the soil will cause water to be sucked through the ceramic cap, exerting pressure which is measured by the transducer. Tensiometers have been used extensively for over a century but are not without limitations. They do not function at very low moisture contents (>100 kPa) as the pores become completely emptied of water. In this case, they will need to be re-saturated. As they are essentially water-filled tubes, they are vulnerable to freezing during very cold conditions. This can damage the porous cup or the interior tubing. Tensiometers need a higher level of maintenance than most sensors, as they can leak, be in poor contact with soil, or exhibit other problems. When servicing, make sure that the manometer or transducer is not loose, and that O-rings are not worn or damaged. You can also reduce leaks by wrapping connections with PFE tape. Don't overtighten! Refill the tensiometer if empty. Make sure that ceramics are not clogged by clay particles prior to installation. You can clean these by soaking in mild chloride solution overnight (Luke and Lantzke, 2004). A comprehensive guide to tensiometry and soil water sampling is provided in Cooper (2016).

Soil moisture content and matric potential are closely related to one another, so both measurement approaches are often used either as proxies or in conjunction with one another. Coupling θ and ψ measurements allows construction of soil water retention curves (SWRCs, sometimes also called soil water characteristic curves), which differ for every soil depending on its texture and structure. These can be used to determine hydraulic parameters via equations such as those developed by Van Genuchten (1980), Mualem (1976), Brooks and Corey (1966), Kosugi (1994), and others, along with values of soil quality (Dexter, 2004). These parameters are used in hydrologic models.

Step-by-Step Installation

Now that you understand the basic concepts of environmental monitoring, let us discuss the elements involved in establishing your monitoring regime. Installation of a sensor array will vary depending on what specific measurements you will be taking, what devices you will use, the location, whether you are using a logger-box or larger housing such as a kiosk, etc. There are some steps however which are similar across most installations.

1) **Assess your site** – When designing your array, you may need to consider the weather, soil conditions, depth, width, or flashiness of a river, exposure to wildlife or livestock and any other defining or influential characteristics of your site. You must consider access to a power supply and the distance and terrain which you will have to cross to access the array in future. Take some preliminary water or soil samples, dig a profile, or set up a temporary station as necessary.

2) **Positioning your array(s)** – Give careful thought to the position of your arrays; the optimum location depends on the reason for your monitoring (Fig. 5.17). For example, continuous water quality monitoring is typically done at the outlet of a catchment or watershed, the point through which all the water from that hydrologic basin drains and exits. As such, measurements taken at this site represent the sum of all hydrologic processes occurring within the wider area, although it may be challenging to disentangle specific drivers. Examples of outlet monitoring designs include Owen et al. (2012), Mellander et al. (2012), McDonald et al. (2015), and many others. Conversely, soil monitoring arrays might best be installed along a hillslope transect, or as part of a plot layout, but their measurements are likely to be indicative of a relatively small

Fig. 5.17 Eddy covariance measurement is used to monitor atmospheric fluxes such as methane and carbon dioxide. Positioning of these and other monitoring arrays is crucial to prevent innacurate measurements. *Source:* Rachael Murphy.

vicinity. Weather stations (Fig. 5.18) must be installed at locations that are representative of the study area and which are away from obstructions such as buildings or trees, with at least a 9 m diameter of open space surrounding (Campbell Scientific, 1997). Maintain the vegetation in this area short. Weather stations should not be installed in shaded areas such as hollows or beneath overhanging vegetation, nor should they be near heat sources. Temperature and humidity sensors should be housed within ventilated enclosures to protect them from ambient radiation. It is common in catchment or watershed-scale studies to maintain more than one rainfall gauge at separate locations. This is for two reasons. First, it provides a backup, as buckets can become obstructed or malfunction and if site visits are irregular or disrupted, this can lead to data gaps. Second, it allows variation in precipitation across the area to be evaluated. This might be critical where there are significant differences in elevation across your study catchment. Pyranometers used for measuring solar radiation should be completely free of shading, artificial light, or reflection. For example, they should not face a lake, as reflected sunlight from the water surface will confound the measurement.

3) **Design** – In designing your array, you need to consider what you need to measure, how many replicates you need, what conditions your sensors and logger need to accommodate, and how you intend to communicate with the array. What you measure should be relatively straightforward and depends on your hypothesis. For example, if you propose that increasing soil moisture content alters microbial respiration rates, you will need volumetric moisture sensors and gas sensors or samplers. You should also consider what sensors or devices you need to provide contextual information such as weather, although this data can sometimes be sourced from nearby stations. In regards to replicates, some sensors (e.g., optical sensors for phosphorus analysis in stream-water) typically are used singly. However, for soil sensors (temperature, volumetric

Fig. 5.18 This weather station includes a rain gauge, anemometer, temperature and humidity sensor, and pyranometer. The station is positioned in the middle of a field to prevent obstruction from trees or buildings. Sturdy fencing protects the station from grazing livestock. *Source:* Sara Vero.

moisture content, matric potential, etc.,) it is more common to install a number of replicates to allow for spatial variability in pore structure and in texture. You may also install at several depths or horizons. Remember to allow sufficient cable to allow sensors to reach from the logger to the target location.

The intended environment needs careful consideration. Often solutions to environmental challenges will be similarly simple, but it is vital to think of these before you get to the field. Take some time to consider; what risks might my array be exposed to in the field? How can I protect it?

Case Study – Protecting a Monitoring Array

Protecting monitoring arrays is often quite simple. For example, I designed some monitoring arrays for the Konza Prairie research facility in Kansas, United States. These arrays needed to monitor soil moisture, temperature, and oxygen at four depths and also incorporated ports for sampling pore water. The challenge was to make the arrays suitable for monitoring during prescribed burns. We wanted the sensors to monitor changes continuously, even while the surface was aflame. All wiring and tubing were contained within PVC piping, and connection to the logger-box was sealed completely. This prevented the soft plastic wire coatings and tubes from exposure to direct heat, which would to melt them. Heat blankets and shields were used to further protect the logger-box during burns, while the sensors monitored changes subsurface.

As part of the design process you should investigate the pros and cons of the various suppliers and individual instruments. Equipment that has been cited in published peer-reviewed studies has the advantage of being "tried and tested." Ask yourself, what have other researchers conducting similar studies used? Don't be afraid to reach out to those researchers for their opinion about field equipment if you are unfamiliar with particular devices. The standard of customer service provided by suppliers may also be important, particularly if support is required after installation. While fraudulent or poor-quality suppliers seem to be thankfully rare, they do exist. Product reviews, FAQs, and online forums are also good ways to gauge the reputation of suppliers.

4) **Get quotes** – You should know how much funding is available to spend on your monitoring equipment and what your institutional guidelines are. Some institutes will allow you to independently select equipment up to a certain price threshold. Even when this is the case, you should obtain quotes from your desired suppliers so that you can review your design and amend to fit your budget if necessary. Other institutes require that you obtain quotes and choose the cheapest or most competitive supplier. This may be less preferable if you are particularly familiar with or prefer devices from a certain supplier, but the intent is to prevent any improper bias toward any company. Check your university or institute guidelines in advance.

5) **Purchase equipment and supplies** – Once you have specified, priced, and selected your equipment, it must be purchased from the supplier. Often you will need to generate a purchase order from your research institution that routes payment directly to the supplier. Ask your finance department how long you should expect this to take as it can vary between different institutions. The actual delivery of equipment will also have a lead time. Always ask the supplier when you can expect the equipment. If there is particular urgency, you may inquire

whether expedited shipping can be arranged, though this usually incurs an additional fee. Keep in mind, if you are purchasing bespoke or custom equipment it may take longer. Don't let this deter you; adjust your scheduling as required. Judge on a case-by-case basis whether it is more important to deploy to the field quickly or to have tailored equipment.

6) **Test all equipment** – Most sensors are tested in the factory to rigorous standards. However, occasional defects can occur, and devices can be damaged in transit. Thankfully these issues are very rare. However, it is better to find out in the lab rather than in the field! You can attach your sensors to the data-logger and battery and check whether they are working in the laboratory or on-site at your facility. Many environmental monitoring companies now produce small, relatively inexpensive handheld loggers which can be connected to various SDI-12 sensors. These loggers offer instant readouts and may incorporate a keyboard and LCD displays so are particularly useful in testing and calibrating sensors, although they also allow spot-checks of sensors in the field.

7) **Calibrate** – Variations in production can cause sensors to deliver slightly different readings. Furthermore, the calibrations conducted in the factory use specific, homogenized reference materials (e.g., pure sands, filtered water, controlled temperatures/humidity, etc.,) which may be significantly different to the soil, rock, and water in which you will deploy them. Although these "out-of-the-box" calibrations are nowadays quite accurate (within a few percentages), it is generally best practice to calibrate to your own substrates and settings.

8) **Write your program** – All sensors and loggers require a program which will control the frequency of measurements, translation of raw data into desired parameters, and communication and/or storage of measured data. Some manufacturers supply user-friendly, "out-of-the-box" software which allow programs to be designed via a highly accessible user interface. One such example is the LoggerNet software offered by Campbell Scientific. The advantage of these proprietary programs is that they allow users with limited coding ability to rapidly and easily prepare loggers for the field. The programs devised using these are suitable for operations of relatively high complexity; however, they may not be entirely as flexible as coding approaches. Free-license code languages such as R, EdLog, and CRBasic are also used by various loggers and may allow more complex program design by skilled programmers. Logger companies often provide custom program designs for a fee or offer customer support via email. There are many very helpful blogs and forums available also, and the online programming community is, in my experience, extremely helpful.

9) **Arrange an installation day** – Your installation day (or days) should be planned in advance so that travel and accommodation to be arranged, but with consideration of weather also. It can be difficult to install in heavy rain or strong wind. Equally, extremely hot conditions at the height of summer can make the heavy work of installation very uncomfortable. As weather forecasts are less reliable further into the future, this may limit how far in advance you can plan. The availability of staff on the proposed day is crucial. While there is no strict rule, it might be wise to schedule your installation early in the working week. That way if an additional day in the field is required you can continue without eating into weekend time. Consider also that you may need time in the laboratory to prepare or store samples immediately after fieldwork and time to travel to and from the site.

10) **The Installation**
Take samples – Soil cores, loose soil, herbage samples, etc., should be taken either prior to starting the installation. Likewise, soil and landscape descriptions should be recorded. Depending on your own expertise and that of your team, it may be best to involve an expert for this element, depending on the level of detail you require. These data will provide contextual

information critical for interpreting your results and also for the site description portion of your Materials and Methods section. Take care to label samples accurately. Installation is demanding and it is easy to mislabel or poorly store samples; take your time or assign this task to a single responsible individual on your team. Regarding site description, this can be written down, but I have had great success using a voice recorder so that I can make observations while I work. Remember to transcribe these observations after fieldwork however!

Excavate – If you are installing sub-surface samplers, you may need to dig a soil pit or trench or augur an access hole. Of course, this will disturb the site somewhat. By laying tarps, you can remove all excavated earth from the surface afterward and minimize damage. Take care to separate each soil horizon as you dig so that soil can be returned to the correct depth.

Secure posts/tripod/frame/housing – Your data-logger may need to be supported or sheltered, and certain devices such as solar panels and air temperature sensors need to be elevated. Similarly, devices may need to be secured to a streambed or at a specified depth within the water column. This can be done by driving a wooden or steel post (Fig. 5.19) and securing your equipment to this structure. Depending on the substrate you may need to construct a cement base. This should be ascertained during a site visit prior to the installation day.

Fig. 5.19 This loggerbox is securely fastened to a steel post. It also safely elevates the datalogger above the maximum water level. *Source:* Brandon Forsythe.

Attach logger-box – Once your tripod or frame is secure, you need to attach your logger-box (Fig. 5.20). This can be done using a bracket and screws, although heavy duty cable zip ties are also useful. The actual data-logger should be securely affixed to the backplate of the logger. It is common to include a humidity indicator card inside the logger-box. These cards are impregnated with a humidity-sensitive chemical (such as cobalt) that changes color in response to moisture. This allows you to determine if the logger-box is leaking or damp, which is potentially harmful to the logger and wiring. You can also place desiccant bags inside the logger-box to help reduce incidental moisture.

Install earth – Properly earthing (or "grounding") the logger is essential to prevent damage to the delicate components during power surges. These can occur due to lightning or electrical discharge from other devices. Loggers include a lug to which you connect the earth wire. Drive a copper or galvanized steel stake (at least 5/8″ diameter) firmly into the soil. It should not be easy to remove. Connect the earth wire from the lug to the stake. Most logger-boxes will have a small port on the underside to pass the wire through. The earth wire should be a heavy gauge, and only as long as is necessary to reach the stake. Longer wires are less capable of rapidly diverting the surge.

Fig. 5.20 This logger box protects the data-logger and sensor wiring from exposure. The black plastic box on the ground is a marine battery box, which is also weatherproof. These sturdy boxes can be purchased from sailing and outdoor shops and are useful for protecting larger batteries. *Source:* Brandon Forsythe.

Install sensors – First connect the wire ends to the appropriate port of the data-logger. This can often be done in the laboratory prior to transport to the field, which saves time and also prevents contamination with dust or moisture. You may have several sensors connected to a single data-logger. Always label the pigtail end of the wire attached to the data-logger with the I.D. of that sensor (Fig. 5.21). This will allow you to identify it later and deal with servicing, malfunctions, or programming.

If using sub-surface sensors (soil or sediment), you usually need to create a guide-hole. This prevents delicate (and often expensive) equipment from becoming bent, chipped, or cracked by striking stones or pebbles. Ideally guide-holes should be snug. You can achieve this by creating a guide that is slightly smaller than the sensor or by allowing a larger hole and filling the space around the sensor with mud made from excavated soil or with kaolinite paste. Take the time to ensure good soil-sensor contact as spaces will allow preferential flow and anomalous readings. For above-ground sensors, make sure that they are securely fixed (e.g., to a strut or bracket). Loose sensors are vulnerable to both error and damage. Always record the position of each sensor in your field notes (e.g., what horizon or depth they are buried at (Fig. 5.22), what GPS location, what height above ground, etc.) It is not very convenient to try and remember later if the sensor was at 10 or 20 cm and you certainly won't want to excavate it and ruin your installation!

Fig. 5.21 Labelling the wires connecting to the datalogger can help if adjustment, repair or replacement of sensors or equipment is required. *Source:* Brandon Forsythe.

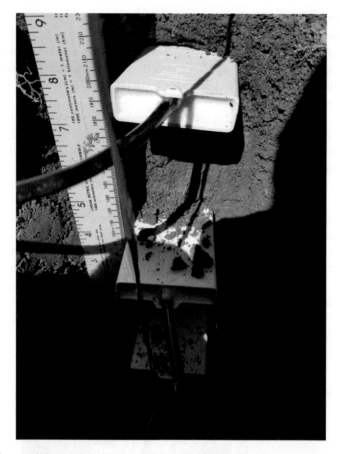

Fig. 5.22 Carefully record the depth, height or position at which you install any sensors, probes or other devices. This is crucial for interpreting the data. *Source:* Brandon Forsythe.

11) **Test all equipment is functioning** – Once your logger and sensors are in place, you must check if they are functioning. Turn on the power and connect to the data-logger using a laptop, tablet, or hand-held field device. You should be able to get a live-feed from the logger. Be aware that immediately after installation, some anomalous readings (noise) may occur. These should settle down as the sensors become established in their position and most modern devices equilibrate to their environment rapidly (within minutes), although certain equipment (such as tensiometers) can take several hours to stabilize (Eijkelkamp, 2008). If there are missing or unintelligible readings, you should check (i) the wiring to the logger, and (ii) the position of the sensor.

12) **Repair and protect the site** – Once everything is working, you will need to tidy wires, refill soil pits, remove debris, etc. Remember that a field site should have minimal damage as a result of your installation. The less intrusive it is, the more likely you are to capture environmental conditions rather than the legacy of your own work. If you installed probes in a soil pit, you will need to refill it by hand. Return all soil to the correct horizon and build the profile gradually. Pack the soil next to the pit face first and work back from there, protecting the sensor cables as you go. Be careful not to cut or tug the cables – this may dislodge or damage your sensors.

Once your soil pit has been refilled you may need to install fencing around the site to prevent interference by livestock or wildlife. Depending on what animals are in the area, this can be relatively lightweight (e.g., single or double-strand electric fencing), sturdy and semi-permanent, (e.g., post-and-rail fencing for cattle) or tall, multiwire fences to exclude deer. These decisions are specific to your scenario but should always be considered prior to installation as damage to your devices may be costly, upsetting, and ultimately can compromise the integrity of your monitoring or treatments. Protecting your site must be a priority, not an afterthought. A word of warning, rodents including rats, mice, squirrels, and other mammals such as rabbits will gnaw exposed wiring. Covering your wires with either a PVC pipe (Fig. 5.23) or with fine mesh will prevent damage by small wildlife.

Fig. 5.23 Covering wires and cables with PVC pipes, as shown here, can prevent trips, damage or gnawing by wildlife or livestock. *Source:* Brandon Forsythe.

13) **Ongoing quality control** – Although it can be time-consuming, checking your data regularly can help you detect trends, alert you to any errors or malfunctions, and allow you to implement calibrations as necessary. Failure to check your data in an ongoing manner can prevent you from detecting and fixing problems, which may undermine the usefulness of your final dataset. Quality control involves checking your dataset for gaps, error messages, or readings that are out of range or unrealistic. These datapoints may need to be removed or can be interpolated from other parameters. It is good practice to screen your data for unusual values outside of expected or reasonable ranges (U.S. Environmental Protection Agency, 2000). These values may require further investigation or comparison to other measurements to determine whether they are accurate or if there is some error or malfunction in the equipment.

Some data-loggers can be programmed to issue alerts via wireless communication when parameters exceed certain ranges. You can specify these ranges so that you are warned of potential equipment malfunction or unusual conditions on-site. It is common for alerts to be activated

Fig. 5.24 Downloading data in the field. *Source:* Brandon Forsythe.

by low battery power. Alerts can also be used to trigger actions in other connected equipment. For example, a flow gauge in a stream can be used to trigger an autosampler at target flow rates.

14) **Scheduled inspection and maintenance** – Similarly to checking your data, you should regularly visit and inspect your installation (Fig. 5.24), even if you are downloading the data remotely via telemetry. Sites require maintenance. Vegetation may need to be trimmed away from the logger. The logger-box should be checked for leaks. Changes to the site may need recording. Batteries need changing. Set a schedule for inspection and maintenance and use a checklist to make sure all the relevant tasks are addressed. This should be considered when planning your sampling regime as some of these maintenance tasks can take place at the same time, but extra time should be allowed.

References

Abatzoglou, J.T. (2011) Development of gridded surface meteorological data for ecological applications and modelling. *International Journal of Climatology* 33(1), 121–131.

Bell, J.E., Palecki, M.E., Baker, C.B., Collins, W.G., Lawrimore, J.H., Leeper, R.D., Hall, M.E., Kochendorfer, J., Meyers, T.P., Wilson, T. and Diamond, H.J. (2013). U.S. Climate Reference Network soil moisture and temperature observations. *Journal of Hydrometeorology* 14(3), 977–988. doi:https://doi.org/10.1175/JHM-D-12-0146.1

Brock, F.V., Crawford, K.C., Elliot, R.L., Cuperus, G.W., Stadler, S.J., Johnson, H.L. and Eilts, M.D. (1995). The Oklahoma Mesonet: A technical overview. *Journal of Atmospheric and Oceanic Technology* 12(1), 5–19. doi:https://doi.org/10.1175/1520-0426(1995)012%3C0005:TOMATO%3E2.0.CO;2

Brooks, R.H. and Corey, A.T. (1966). Properties of porous media affecting fluid flow. *Journal of the Irrigation and Drainage Division* 92(IR2), 61–88.

Campbell Scientific. (1997). Weather station siting and installation tools. App. Note Code 4-S. Logan, UT: Campbell Scientific.

Campbell Scientific. (2011). Power supplies. App. Note Code 5-F, Revision 11. Logan, UT: Campbell Scientific.

Cooper, J.D. (2016). *Soil Water Measurement: A Practical Handbook*. London, U.K.: Wiley Blackwell.

Dexter, A.R. 2004. Soil physical quality: Part I. Theory, effects of soil texture, density, and organic matter, and effects on root growth. *Geoderma* 120(3–4), 201–214.

Eijkelkamp Agrisearch Equipment. (2008). *Operating Instructions: 14.04.03 Tensiometer and 14.04.04 Jet Fill Tensiometer*. Morrisville, NC: Eijkelkamp.

European Parliament, Council of the European Union. (2012). Directive 2012/19/EU of the European Parliament and of the Council of 4 July 2012 on waste electrical and electronic equipment (WEEE) Text with EEA relevance. COD 2008/0241, EEA relevance. Brussels, Belgium: Publications Office of the European Union.

van Genuchten, M.T. (1980). A closed-form equation for predicting the hydraulic conductivity of unsaturated soils. *Soil Science Society of America Journal* 44, 892–898. doi:https://doi.org/10.2136/sssaj1980.03615995004400050002x

Kalbus, E., Reinstorf, F., Schirmer, M. (2006). Measuring methods for groundwater-surface water interactions: a review. *Hydrology and Earth System Sciences Discussions* 10(6), 873–887.

Kinouchi, T., Yagi, H. and Miyamoto, M. (2007). Increase in stream temperature related to anthropogenic heat input from urban wastewater. *Journal of Hydrology* 335, 78–88.

Kosugi, K. (1994). Three-parameter lognormal distribution model for soil water retention. *Water Resources Research* 30(4), 891–901.

Luke, G. and Lantzke, N. 2004. *Tensiometers – Preparation and Installation*. Washington, D.C.: Department of Agriculture Farmnote, USDA.

McDonald, K.J., Reynolds, B. and Reddy, K.J. (2015). Intrinsic properties of cupric oxide nanoparticles enable effective filtration of arsenic from water. *Scientific Reports* 5, 11110.

Mellander, P.-E., Melland, A.R., Jordan, P., Wall, D.P., Murphy, P.N.C. and Shortle, G. (2012). Quantifying nutrient transfer pathways in agricultural catchments using high temporal resolution data. *Environmental Science and Policy*, 24, 44–57. doi:https://doi.org/10.1016/j.envsci.2012.06.004.

Mellander, P-E., Jordan, P., Shore, M., Melland, A.R. and Shortle, G. (2015). Flow paths and phosphorus transfer pathways in two agricultural streams with contrasting flow controls. *Hydrological Processes* 29(16), 3504–3518.

Mualem, Y. 1976. New Model for predicting hydraulic conductivity of unsaturated porous-media. *Water Resources Research* 12, 513–522. doi:10.1029/WR012i003p00513

Owen, G.J., Perks, M.T., McW, C., Benskin, H., Wilkinson, M.E., Jonczyk, J. and Quinn, P.F. (2012). Monitoring agricultural diffuse pollution through a dense monitoring network in the River Eden Demonstration Test Catchment, Cumbria, UK.

Panero, S., Romoli, C., Achilli, M., Cardarelli, E. and Scrosati, B. (1995). Impact of household batteries in landfills. *Journal of Power Sources* 57(1–2), 9–12.

Petropoulos, G.P. and McCalmont, J.P. (2017). An operational in situ soil moisture & soil temperature monitoring network for West Wales, UK: the WSMN network. *Sensors* 17(7), 1481. doi:https://doi.org/10.3390/s17071481

Pica, J, (ed.), (2018). *Requirements and standards for NWS climate observations. National Weather Service Instruction 10-1302.* Silver Spring, MD: National Weather Service.

Sherriff, S.C., Rowan, J.S., Fenton, O., Jordan, P., Melland, A.R., Mellander, P-E. and O'hUallacháin, D. (2016). Storm event suspended sediment-discharge hysteresis and controls in agricultural watersheds: Implications for watershed scale sediment management. *Environmental Science and Technology* 50(4), 1769–1778.

U.S. Environmental Protection Agency. (2000). *Meteorological Monitoring Guidance for Regulatory Modeling Applications.* Washington, D.C.: USEPA.

6

Soil Techniques

Soil Sampling Equipment

The precise equipment which you will use in the field will vary depending on what techniques you happen to be using. However, as discussed in Chapters 3 and 4, it is best to be prepared and have a variety of tools at your disposal. Here is a list of common soil sampling tools you are likely to need. Remember, when it comes to fieldwork, your equipment should be tailored to your needs, so add to or edit this list as you see fit.

Soil sampling equipment		
Shovel	Rubber mallet	Knife
Auger	Lump hammer	Permanent markers
Measuring tape (metric)	Bucket	Notepad
Trowel	Munsell chart	Pens
Bulk density rings and caps	Sample bags (various sizes)	

Other equipment such as infiltrometers (for measuring soil conductivity) or penetrometers (for measuring strength or indicating compaction) may be useful, depending on your objectives.

Soil Survey

There are many reasons to conduct soil surveys, so the approach to sampling and types of samples differ.

Nutrient indices/balances – Surveys are often conducted relatively close to the soil surface to determine the availably of nutrients required for crop growth, typically within 0–20 cm (or 0–8 in.). However, this will vary depending on the purpose of your study and the type of crop or land use. Where the objective is to evaluate available nutrients (either for crop use or environmental loss), the depth of sampling should reflect rooting depth and potential stratification of nutrients. This will vary depending on the nutrient (e.g., phosphorus, Baker et al. 2017) and the type of tillage (Crozier et al. 1999).

Fieldwork Ready: An Introductory Guide to Field Research for Agriculture, Environment, and Soil Scientists, First Edition. Sara E. Vero.
© 2021 American Society of Agronomy, Inc., Crop Science Society of America, Inc., and Soil Science Society of America, Inc. Published 2021 by John Wiley & Sons, Inc.
doi:10.2134/fieldwork.c6

Where sampling for nutrient concentrations is required by legislation (as in the United Kingdom, for example), mandatory depths of sampling will be specified (e.g., 75 mm for grassland, 150 mm for arable land). If your research is intended to be relevant within a particular legislative context such as the above, you should conduct your sampling in accordance with the legislated standards.

The absolute values of nutrient concentrations may be allocated to index systems. These systems assign a number or ranking to soil nutrient concentration ranges, with each number specifying a level of nutrient availability or risk. Index systems may rely solely on concentration (e.g., the Irish and UK phosphorus indices) or may incorporate other factors such as runoff class, tillage system, denitrification potential, proximity to a receptor (Delgado et al. 2006). Index systems are a helpful tool for advisory and/or extension to ensure fertilizer recommendations are accurately prescribed to satisfy crop requirements at field scale and, in many regions, are also used to ascertain environmental risks arising from excess nutrients (e.g., United Kingdom). Sharpley et al. (2003) reported that 47 US states have adopted phosphorus index systems, although the specific soil test, ranges, and calculation method varies from state to state.

Soil test results can also be used to investigate potential metal toxicities and increasingly, are used as an aid to precision management of irrigation, seeding rates, and nutrient application. Each field or paddock should be sampled individually, and large fields should be divided into smaller units. Areas which have different management, soil type, or drainage should be sampled separately. Sampling should not be conducted within 3 months of fertilizer or lime application, as these will skew the results. The depth of sampling varies depending on soil type, nutrient, crop, and on national standard test depths. This should be determined ahead of time based on the relevant literature, or multiple depths may be evaluated. The frequency of nutrient sampling for preparing agronomic recommendations varies. In the United Kingdom and Ireland, testing is required every 3–4 yr. Within the United States, sampling frequencies are prescribed on a state-by-state basis, often suggesting at least every 2–3 yr. However, these are considered to be the minimum, and annual sampling is encouraged for agronomic planning.

The sampling frequency used in your study should reflect your objectives, the nutrient in question, and the duration of your study.

Sampling procedure is as follows:

1) Avoid sampling when the soil is extremely dry or hard as it will be difficult to drive in the corer. Also avoid extremely wet conditions, as the corer is likely to become clogged.

2) Push the soil corer into the soil to the full depth of the funnel by pressing on the handles and using the foot-step to help you drive it into the soil.

3) Repeat at intervals until collection hopper is full. See Fig. 2.15 for sampling patterns.

4) Homogenize the samples by mixing in a bucket and store in a labeled sample bag or box.

5) Soil samples are commonly dried before analysis or storage. The method of drying depends on the analysis to be conducted as mineralization or decomposition can occur. However, it is common to air dry and sieve to <2 mm prior to most chemical analyses (Soil Survey Staff 2014). This is not suitable for samples undergoing biological analyses in which field-moist samples are usually preferred. Air drying is done by spreading the sample on a pre-weighed tray and drying in an oven at 30–35 °C or in an ambient room (c. 27 °C) for 5–7 d, until weight remains constant. The samples should be crumbled to expose a greater surface area. This will encourage evaporation and even drying. You may need to turn the samples every few days. Always consult the standard operating procedure (SOP) for your specific analyses before drying soil samples.

6) Storage of dry samples should be done by packing in airtight bags, boxes, or jars and storing in cool, dark conditions. Be aware that some changes may occur during prolonged storage. Always label stored samples with the date of archival.

7) Ideally, laboratory analysis for nutrient concentrations should be performed shortly after sampling to prevent any degradation or alterations within the sample during storage. The effect of storage varies depending on the property of interest (Sun et al. 2015).

Bulk density – (ρ_b) is a key property that influences the infiltration capacity, strength, penetration resistance, and quality of the soil. It can be altered by soil management or compaction by vehicles or livestock and is sometimes used as a proxy for overall soil health or quality. Typically, ρ_b varies spatially across an area and increases with depth through the soil profile. Sampling for ρ_b is conducted as follows:

1) Remove vegetation from the soil surface.
2) Take a steel ring (5 cm Ø) (ideally with a sharp edge). Press the sharp edge into the soil surface.
3) Place a wooden block on top of the ring and use a rubber mallet to drive the ring into the soil until the block is flush with the soil surface. Alternatively, you can use a pipe or driving ring with the same diameter as the ring in place of the block. This will prevent accidental compaction and allow you to drive the ring to just below the surface (Fig. 6.1).
4) Extract the core using a trowel.
5) Trim excess soil from both ends of the ring using a sharp knife.

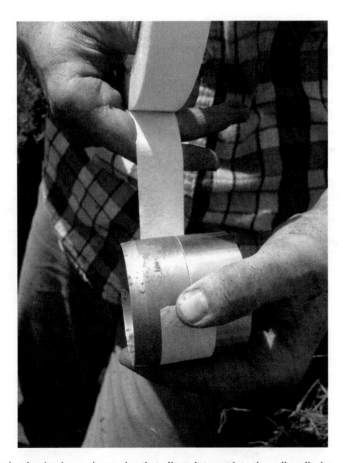

Fig. 6.1 This bulk density ring has a sharp edge that allows it to cut into the soil easily. A second ring with no cutting edge is being taped to it. This allows the sampling ring to be hammered into the soil without compacting the surface. It is also common to use a wooden block. *Source:* Sara Vero.

6) Label rings and store securely so that soil is not knocked out during transport. You can purchase caps that fit snugly on the top and bottom of the rings and prevent soil loss. Alternatively, you can wrap the samples in cling film or simply bag them if you can keep them flat during transit.

7) If you are sampling from a soil profile, you should take cores from each horizon or from specific depths (e.g., every 10 cm) below the surface (ideally >3 reps per horizon). Targeting specific depths may fit your experimental design; in other words, you might be interested in the chemical or physical properties at a certain depth (e.g., 20 cm). However, if you are characterizing a soil pit, taking samples every horizon is more typical. Ideally, you should take each core in a vertical orientation and then dig down a "step" to the next horizon and take the next row of samples. If it is not possible to alter the pit like this (i.e., for safety reasons), you may drive the rings horizontally into the soil profile.

8) Weigh your sample on returning to the laboratory and dry at 105 °C for 48 h. Reweigh the sample. Bulk density is then calculated as follows:

$$P_d = \text{Dry weight/Volume of sampling ring}$$

You can also calculate the volumetric water content (θ) at the time of sampling as follows:

$$\theta = (\text{Wet weight} - \text{Dry weight})/\text{Volume of sampling ring}$$

Soil profile description – The study of soil types and their characterization is known as "pedology". This science allows insights into the formation and behavior of a soil, its suitability for various cropping systems, its hydrology, and its structural stability. For many field studies, it is common to consult a professional pedologist who can provide a detailed and accurate description of the soil profile using the appropriate classification system (Unified Soil Classification System, World Reference Base for Soil Systems, USDA Soil Taxonomy) (Fig. 6.2). However, anyone can dig a soil pit and gain substantial information about their study site. Most

Fig. 6.2 A colourful soil profile. *Source:* Jaclyn Fiola.

agricultural or engineering universities will offer modules on pedology, which is highly recommended for anyone working in environmental science. In addition, manuals on detailed soil profile description such as the NRCS Survey Manual (Soil Science Division Staff 2017) are invaluable and will help you adhere to national conventions. An illustrated guide to soil taxonomy is also available on the NRCS website.

Here are the basic steps for recording a simple soil profile (such as for background or contextual knowledge):

1) As discussed previously, you may need to consult local utilities suppliers or residents before excavating soil pits to avoid damaging buried cables or pipes. This may be particularly important in areas near housing, bridges, ditches, or junctions.

2) Record the site description before you have disturbed it in any way. You must record the date, who is conducting the survey, the weather at the time, elevation, hillslope position (summit, shoulder, backslope, footslope, toeslope), aspect, and dominant vegetation. Take note of key landscape features such as outcrops, floodplains, terraces, changes in vegetation, etc. It is useful to indicate drainage class, but be careful! Sites may appear to have better or worse drainage than is typically the case depending on antecedent weather conditions. Once you have excavated the pit, there will be indicators of the more long-term drainage status, so do not make assumptions based on what you see above ground alone. Always record latitude and longitude, and if using GPS devices, you should tag the location. Take photos. If your site is on agricultural land, the farmer may be able to provide detailed information on its general behavior and characteristics.

3) Excavate your pit. This can be done with a digger or backhoe, or by hand. Pits are frequently dug to either bedrock or groundwater, but this may not be possible depending on the depth of these zones at your particular site. Loess soils, for example, may be several meters deep, which is beyond the practical depth of excavation in many cases. Safety is crucial when digging pits. Deep, unsupported pits are more vulnerable to collapse or slumping, particularly if they have weak soil structure. If you are working in deep pits, you may need to establish buttresses to prevent this. OSHA specify that pits deeper than 1.25 m must be shored up. Wet conditions may also reduce pit stability. It is not advisable to enter pits which are deeper than 1–2 m (Logsden et al. 2008). Remember, a collapsing pit can cause crushing, burial, or asphyxiation. Both injuries and, sadly, fatalities have occurred in the research field as well as in construction, farming and other industries.

4) It is a good idea to remove the topsoil by cutting sods first. These will provide intact units held together by vegetation, which will help the surface of the pit return to its original condition.

5) Record the base of the profile. Record the parent material; bedrock, glacial drift, alluvium, etc., and take a sample. If you have reached bedrock, record the type of rock.

6) If groundwater is reached or begins to seep into the soil pit, record the depth at which this occurs.

7) Stake a measuring tape to the soil surface and unroll to the bottom of the pit. Tapes for soil profiling are usually heavy-duty, wide, and clearly marked. Record the depth of the profile and the depth of each horizon. Take a photo. Always include the tape for reference. When taking photos of specific horizons or observations within the profile, use a ruler for scale. It is common to use a pen or trowel as an alternative if you do not have one available.

8) Identify horizons by finding the boundaries at which soil properties change (Fig. 6.3). Color is often the clearest indicator of a change in horizon, although texture, density, and structure are also key. Designate each master horizon in accordance with your selected classification system. Master horizons indicate the dominant characteristics of that layer and are indicated by capital

Fig. 6.3 Cupcake tins are useful for separating soil samples from different horizons while hand texturing. *Source:* Jaclyn Fiola.

letters. A secondary feature of that horizon can be indicated by a lower-case letter. Some differences exist between classification systems, but the USDA subsurface designations are as follows:

O – If present, by organic material at various stages of decay and weathering. This horizon will have low mineral content and will often be the upper layer of the soil. However, inversion of the profile or repeated deposition of new material can result in buried organic horizons.

A – These are true mineral horizons occurring at the soil surface or just below the organic layer. A horizons are closely influenced by surface weathering processes, anthropogenic actions such as cultivation and integration of decaying organic matter (humus) (Fig. 6.4). Horizons are typically dark in color. Plowed A horizons can be indicated by lower-case "p," e.g., Ap.

E These are also mineral horizons but exhibit loss of metals such as iron and aluminum, and of organic matter, clay, or salts through repeated or prolonged leaching (eluviation). These horizons can be identified by pale or grayish color and will always overlie a B horizon.

B – These horizons form the subsoil. They are mineral soils in which Fe, Al, humus, clay, etc., have accumulated from layers higher in the profile. B horizons will have discernible structure. Secondary descriptions for these horizons include leached organic matter (h), accumulated clay (t), accumulated iron (Fe) and aluminum (Al) leached from overlying soil (s) or Fe- or Al-oxides (o). Thin, dense B horizons can occur as a result of major accumulation of metals (iron-pan) or compaction (plow-pan) and will restrict or redirect water flow within the profile.

Fig. 6.4 This soil from the A horizon shows friable structure and has been penetrated by plant roots. This horizon is strongly influenced by cultivation and weathering. *Source:* Sara Vero.

C – These horizons closely resemble the underlying parent material and/or bedrock and have experienced little weathering. Their color will usually closely resemble that of the underlying rock.

R – Bedrock. You are unlikely to dig far into this by hand! Note the type of rock (granite, limestone, basalt, and so on) and its structure – is it fractured?

Other sub-descriptions for soil horizons include "b" for a buried horizon, "g" for gleying (reduction of iron due to prolonged saturation, indicated by gray or mottled soils, (Fig. 6.5)), "ss" for slickensides (shiny, polished surfaces that occur in cracks within soils having high clay contents) and "r" for weathered bedrock. Within the profile, you may have multiple distinct master horizons, or more than one horizon of each type (e.g., two or more B horizons). Where there is a gradual transition between horizons rather than a distinct boundary, they may be indicated by two uppercase identifiers, for example, AB.

You should also record the presence of any anthropogenic artifacts and the depths at which they occur. Human legacy may also present within the soil profile as compaction or as a plow-pan, a dense layer of soil occurring just below plowing depth as a result of repeated tillage and vehicle trafficking, particularly if cultivation was conducted during wet conditions. You may also observe iron-pan horizons. These are typically thin horizons which are reddish in color and result from accumulation of iron oxides in acid soils. If you encounter such horizons, record their depth and thickness.

Fig. 6.5 Mottled colors such as those seen here indicate reduction of iron which occurs due to alternating soil saturation and aeration. *Source:* Jaclyn Fiola.

If you wish to assign the profile to a soil series using the World Reference Base or USDA soil taxonomies, you must identify and record the diagnostic horizons. Diagnostic horizons are those layers that dictate what soil series it is and indicate its behavior. Failure to observe key horizons is problematic and makes characterization more challenging so it is important to excavate to a sufficient depth, usually ≥ 1 m. Unlike master horizons which are recorded using the letter system described above, diagnostic horizons have titles such as argillic, spodic, or mollic. These indicate the processes involved in the formation of that soil. To identify diagnostic horizons and assign a profile to the correct order and series you will need to be familiar with the classification system and have a taxonomic key available. Diagnostic classification is somewhat more complex than profile description using only master characteristics, and training in soil characterization is recommended. Many universities have soil judging teams which will help you develop these skills, or may even offer their assistance in the field.

9) Record the boundary of each horizon. Is it smooth, wavy, irregular, or broken?
10) Record the texture of each horizon. Soil texture is the distribution of sand, silt, and clay particles (Figs. 6.6 and 6.7). There are various textural classification systems (so always record which system you are using). These systems prescribe slightly different particle size ranges (Table 6.1), so clearly recording the system you are using will allow accurate interpretation of your results. While training and experience are crucial, basic hand-texturing can be conducted as follows (Figs. 6.8–6.15). A step-by-step guide to hand-texturing is provided in Fig. 6.9.

Fig. 6.6 Hand texturing can be used to determine soil textural class (see flowchart). This soil has a large proportion of sand particles and a granular structure. *Source:* Jaclyn Fiola.

Fig. 6.7 Grey coloration indicates prolonged periods of saturation. This soil also has a slightly platy structure, indicated by flat, layered peds. *Source:* Jaclyn Fiola.

Table 6.1 Soil particle size classes according to United States Department of Agriculture (USDA), World Reference Base (WRB) and International Society of Soil Science (ISSS) classification systems.

Classification system	Grain size (mm)							
	Clay	Silt	Very fine sand	Fine sand	Medium sand	Coarse sand	Very coarse sand	Gravel
USDA	<0.002	0.002–0.05	0.05–0.10	0.10–0.25	0.25–0.50	0.50–1.00	1–2	>2
WRB	<0.002	0.002–0.063	0.063–0.125	0.125–0.20	0.20–0.63	0.63–1.25	1.25–2	>2
ISSS	<0.002	0.002–0.02		0.02–0.20		0.20–2		>2

Fig. 6.8 Form a ribbon with the damp soil using your thumb. How long can you form a ribbon before it breaks under its own weight? *Source:* Jaclyn Fiola.

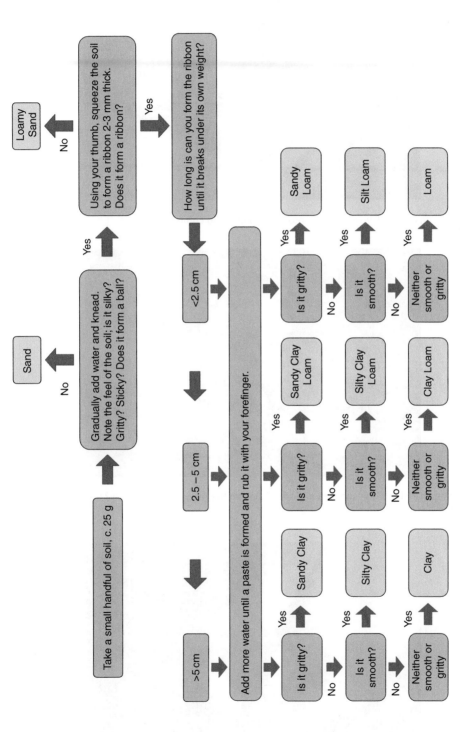

Fig. 6.9 Simple guide to identify soil texture by hand, adapted from Thien et al. (1979). Note that this only determines textural class, although experience will help more accurate assessment.

Fig. 6.10 Forming a high clay soil sample into a ball. *Source:* Bo Collins.

Fig. 6.11 Take a handful of soil, moisten and knead. Record the feel: is it gritty? Silky? Sticky? *Source:* Jaclyn Fiola.

Fig. 6.12 Test for a smooth or gritty feeling. *Source:* Jaclyn Fiola.

Fig. 6.13 Silty soil has a smooth, silky feel when damp. *Source:* Sara Vero

Fig. 6.14 Practice is the best way to become comfortable with hand texturing. Undergraduate modules, field days, soil judging teams or workshops are opportunities to build on these skills. Don't be afraid to ask more experienced colleagues for advice. *Source:* Sara Vero.

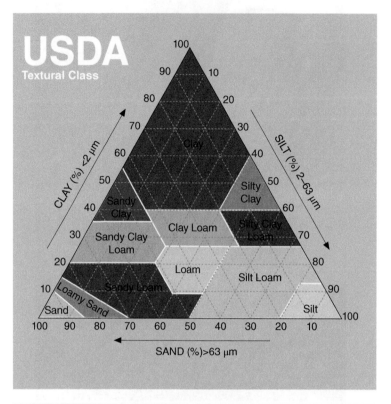

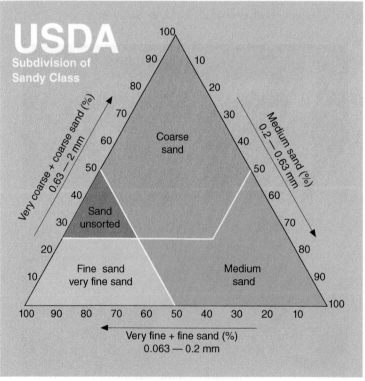

Fig. 6.15 USDA soil textural triangle. *Source:* Karen Brey.

Unified Soil
Classification System

Coarse-grained soils

Gravels More than 50% of coarse fraction larger than No. 4 sieve size

Clean gravels (less than 5% fines)

GW	Well-graded gravels, gravel-sand mixtures, little or no fines
GP	Poorly-graded gravels, gravel-sand mixtures, little or no fines

Gravels with fines (more than 12% fines)

GM	Silty gravels, gravel-sand-silt mixtures
GC	Clayey gravels, gravel-sand-clay mixtures

Sands 50% or more of coarse fraction smaller than No. 4 sieve size

Clean sands (less than 5% fines)

SW	Well-graded sands, gravelly sands, little or no fines
SP	Poorly graded sands, gravelly sands, little or no fines

Sands with fines (more than 12% fines)

SM	Silty sands, sand-silt mixtures
SC	Clayey sands, sand-clay mixtures

Fine-grained soils

Silt and Clays Liquid limit less than 50%

ML	Inorganic silts and very fine sands, rock flour, silty of clayey fine sands or clayey silts with slight plasticity
CL	Inorganic clays of low to medium plasticity, gravelly clays, sandy clays, silty clays, lean clays
OL	Organic silts and organic silty clays of low plasticity

Silt and Clays Liquid limit 50% or greater

MH	Inorganic silts, micaceous or diatomaceous fine sandy or silty soils, elastic silts
CH	Inorganic clays of high plasticity, fat clays
OH	Organic clays of medium to high plasticity, organic silts

Highly organic soils

PT	Peat and other highly organic soils

Primary Component

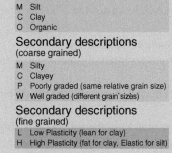

G	Gravel
S	Sand
M	Silt
C	Clay
O	Organic

Secondary descriptions
(coarse grained)

M	Silty
C	Clayey
P	Poorly graded (same relative grain size)
W	Well graded (different grain sizes)

Secondary descriptions
(fine grained)

L	Low Plasticity (lean for clay)
H	High Plasticity (fat for clay, Elastic for silt)

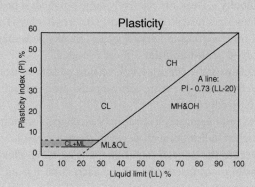

Fig. 6.15 (Continued)

Fig. 6.16 Examining the structure of the horizons. *Source:* Jaclyn Fiola.

11) Record the structure of each horizon (Fig. 6.16). Structure is the arrangement and aggregation of soil particles. Totally loose particles exhibiting no cohesion (such as pure sand) are unaggregated. "Massive" soils also occur which contain no clear structural units and instead form one continuous, unaggregated mass. However, most soils will exhibit at least some degree of aggregation. Each aggregate or clump of particles is a "ped" and it is the characteristics of these peds which indicate the structure.

Granular – Each ped is roughly <5 cm in diameter and is roughly spherical or crumb-like. Granular structure is more commonly encountered in upper horizons (A or B) that are influenced by plant rooting, cultivation, and turbation by vertebrates and macroinvertebrates. These peds may be held together by humus. A granular structure is usually desirable in cultivated soils as it allows root penetration and adequate water movement.

Platy – These are flat peds that are aligned horizontally. Platy horizons can impede water infiltration to lower horizons and may cause lateral flow.

Blocky – These are relatively large polyhedral peds which fit together with smooth ped faces. This structure most frequently occurs in clay-rich soils as a result of shrinking or swelling during wetting and drying cycles. Blocky peds may be categorized as **angular** – having sharp edges, or **sub-angular** – more rounded.

Columnar – These are tall, vertical peds that may have a smooth, rounded top.

Prismatic – These are vertical peds, similar to columnar structures but with smooth faces and sharp edges.

12) Record the presence of rocks (size, shape, location, amount) in each horizon. Also record biological indicators such as worm channels, roots.

13) Record the color of each horizon (Fig. 6.17). This is done by matching the soil to color chips within a Munsell color chart. Each color has an identifier which includes **hue**, **value**, and **chroma**. Hue is the dominant color; how red (R) to yellow (Y) the soil is. Value indicates how light or dark a color is, from 2 to 8 (the actual scale runs from 0 to 10, that is, absolute black to

Fig. 6.17 Some colorful soils! *Source:* Karen Vaughan.

Fig. 6.18 Munsell charts are used to identify the exact color of soil samples according to hue, value and chroma. *Source:* Jaclyn Fiola.

absolute white, although these are not observed in soils). Chroma indicates the strength of the color (0–10, typically, 1–8 in most soils), with higher numbers being more vibrant. You read a Munsell chart by first recording the hue from the top of the appropriate page, then matching the soil to the appropriate chip and recording first the value from the Y-axis and then chroma from the X axis (Fig. 6.18). The color of a soil might therefore be recorded as 7.5 yr 6/3.

It takes experience to become accurate with this. It is common for soil within a single horizon to have a dominant (matrix) color and one or several less-dominant mottles. These mottles may be evidence of leaching (transport of minerals from overlying horizons in the profile) or redox processes. If performing a more detailed pit description, you may want to break open soil peds and record the colors of the inner faces.

Color is a good indicator of redoximorphic features (alterations in the soil caused by wetting and drying cycles and exposure to oxygen). These features tell us about the movement of minerals such as iron through the profile and the moisture regime. Redoximorphic features include mottles (streaks or patches of color – often red or orange), nodules and concretions (hard masses of precipitated iron or manganese), pore linings (soil pores which have become coated with a thin layer of iron or manganese), and zones of depletion (soil which is faded or dull in color as a result of leaching of pigment).

14) Take a loose, bagged sample from each horizon for lab analysis. You may also take intact cores so that bulk density can be measured.

15) Backfill the pit. Try to return soil to the depth from which it was excavated, if possible. This can be made easier by separating the soil from each horizon as best you can during excavation. It is rarely possible to return a pit perfectly to its original state, however, taking time to backfill carefully can help minimize subsidence thereafter. If you removed large rocks, roots, or other debris, these can be returned to help the structural stability of the pit. Finally, replace the sods on top and press down. Take a photo of the final site.

Based on the description you may be able to assign the soil profile to a particular series. A soil series is a group of soils having similar origin and characteristics, such as similar horizons. The concept allows soils with the same traits to be mapped. Soil series are typically named after the locations at which they were first described, but a series could be identified at locations distant from that original site. You can identify which series your soil profile is by matching it to descriptions given in national databases and systems (see Table 6.2 for some resources).

There are excellent resources (some listed in Table 6.2) to help you become familiar with soil description, but there is no substitute for training and experience. Many universities take part in competitive soil judging. Consider joining your campus team or seek out the advice of these skilled individuals if you are new to the method.

Soil augering – Augering is often conducted to validate soil maps (i.e., does the soil at that location actually correspond to the series indicated on a map) or to provide a general indicator of soil characteristics without performing a full soil pit. Augering is much quicker and less labor intensive and can be used in conjunction with pits to provide a more comprehensive assessment of the area. Essentially, a soil auger is a gouge or corkscrew-type device with a T-shaped handle on top. It is twisted directly into the ground and then pulled upward, taking a tall core of soil (Fig. 6.19). This core can then be analyzed for depth, color, texture, etc., just like

Table 6.2 Some resources for soil taxonomy and classification.

Country	Agency	Soil survey database
United States of America	National Cooperative Soil Survey	https://www.nrcs.usda.gov/wps/portal/nrcs/main/soils/survey
United Kingdom	Land Information System	www.landis.org.uk
Ireland	Irish Soil Information System	http://gis.teagasc.ie/soils
Europe	European Soil Database and Soil Properties	https://esdac.jrc.ec.europa.eu/resource-type/european-soil-database-soil-properties
Australia	Australian Soil Resource Information System	http://www.asris.csiro.au
New Zealand	New Zealand Soils	www.nzsoils.org.nz

Fig. 6.19 Augering involves twisting of driving a gouge or corkscrew sampler into the soil and withdrawing a core. Extension rods can be used to sample from greater depths, although this depends on the resistance of the soil and it's stone content. *Source:* Sara Vero.

horizons in a pit. There are several varieties of hand-auger. Select yours depending on the site and purpose. Augers can be fitted with extension rods to allow deeper excavation.

Gouge auger – Suitable for sampling relatively soft profiles to a depth of 0.5–1 m, these augers have an open face and are pushed directly down into the soil before extraction (Fig. 6.20).

Fig. 6.20 A gouge auger with a tapered cutting edge. *Source:* Sara Vero.

Screw auger – These are twisted into the soil profile before being pulled upward. They have a bladed sampling head which allows them to penetrate deeper or more dense, stony, or hard soils. Various screw heads exist, such as:

Dutch auger – These have a cutting edge and an open head to allow collected soil to be easily emptied. These augers are good for penetrating densely rooted, wet, or clay-rich soils.

Bucket auger – These usually have a closed head (the bucket) and two blades. Bucket augers are used in a variety of applications and soil types.

Spiral auger – These resemble a corkscrew and extract a wholly disturbed soil sample. They are typically used for very hard or impenetrable soils.

A chief limitation of hand augering is that you are limited in the depth from which you can effectively extract. This can be overcome in part by extension kits. Using these, you screw the auger to its complete depth, attach the extension, and then continue to insert farther into the soil. When extracting an auger, you should push the handle directly upward, driving with your legs. If deeper or more challenging conditions are encountered (e.g., compacted or rocky layers) you may need to use a mechanical corer (Fig. 6.21). These work by vibrating the auger into the soil and use a lever system to extract the core. Always wear appropriate personal protective equipment (PPE) when operating these, including steel-capped boots and ear protection. Finally, vehicle-mounted augers offer a solution for very difficult conditions or when particularly large cores are required. Some major advantages of augering are that cores can be sectioned (Figs. 6.22 and 6.23) so that they can be analyzed individually in the laboratory (Fig. 6.24) and also, depending on your equipment, intact cores can be removed and used for lysimeter studies under controlled conditions (Figs. 6.23 and 6.24).

Fig. 6.21 Mechancal corers use a motor to drive the sampler into the soil. This allows sampling from greater depths than is usually possible with hand-operated augers. *Sources:* Jaclyn Fiola and Karen Vaughan.

Fig. 6.22 A soil core divided into sub-sections for analysis. *Source:* Jaclyn Fiola.

Fig. 6.23 Soil cores should be measured and can be subdivided into sections corresponding to different depths or horizons. These researchers have laid the extracted core on a tarp while characterising and sampling each section. *Source:* Jaclyn Fiola.

Fig. 6.24 PVC pipes can be used to store and transport soil cores for analysis in the laboratory. *Source:* Jaclyn Fiola.

Pore-Water Sampling

Sampling of soil pore water is increasingly popular in assessments of nutrient and contaminant transport. Pore-water is the moisture held in the voids and cracks between solid soil particles. At any time, these pores are filled with liquid and gas in varying proportions, which influences microbial activity, soil trafficability, and the movement of solutes both vertically and laterally through the subsurface. In the root zone, the contents of pore water will indicate the availability of essential nutrients to the growing crop and the presence of solutes.

There are two primary approaches to sampling the pore water: destructive and nondestructive. Destructive sampling entails taking a bulk soil sample, that is, by excavating a block of soil from the target depth. The liquid portion of that soil must then be drained out and collected. This approach has major limitations; it is not repeatable at that site and so can provide only snapshot data and not a time series, and the sample may be vulnerable to evaporation or other losses during transport and processing which can influence results.

It is generally preferable to extract a liquid sample from the intact soil profile. This can be done using porous suction cups attached to flexible tubing routed to the surface. Suction can be applied using pumps or syringes (Fig. 6.25) to extract a sample through the cup. Very small cups measuring only a few millimeters in diameter are now available. These can be inserted into a soil profile or into potted samples or soil columns. When using suction cups of any size, you must be cognizant of the diameter of pores. The size of these pores will dictate what size of solute or suspension can be extracted from the soil. This is particularly important if you are attempting to extract suspended colloids which are often too large for extraction using this method. Pesticide and herbicide molecules may also be too large. You can test this quite easily in the laboratory by preparing a liquid suspension or solution containing your targeted chemical, and then extracting a sample using your suction cups. Analysis of this extract will indicate whether your target has been successfully taken in via the porous cup.

Fig. 6.25 Suction is applied to soil pore water samplers using a pump or syringes (shown here). This method allows water held within the soil matrix to be extracted for laboratory analysis. *Source:* Sara Vero.

Whether you use suction cups or smaller samplers, they should be saturated before installation. This can be done by soaking them overnight in water and then drawing a syringe or two of water through the sampler. By doing so you can ensure that a continuum of water is available for transport.

There are several approaches to installing suction cups. The simplest approach is to augur a hole to your target depth within the soil and to insert your sampler, using a kaolinite or soil slurry to ensure good soil sampler contact. You should not install the sampler vertically (at 90° to the soil surface) as this can encourage preferential flow along the tube. Keep in mind that if you install at an angle the sampler will not reach so deeply into the soil. You can calculate what depth it is located by simple trigonometry.

A potential limitation of installing from the surface is that you cannot be certain what horizon you are sampling from. An alternative approach, particularly when using smaller samplers, is to install them laterally into a soil profile and then to route the tubing to the soil surface via an access pipe. The pit or hole can then be backfilled. This allows you to precisely target certain horizons. Small samplers can be quite fragile so a pilot hole which is slightly narrower than the sampler should be made using a drill or simply a sharp implement such as a Phillips screwdriver. Make a slurry or paste using kaolinite or soil from the profile and water and coat the sampler prior to inserting gradually into the pilot hole (Fig. 6.26). You can install the sampler to a target depth by making the pilot hole deeper and by backfilling using soil slurry around the sampler tubing. You can simply backfill the pit leaving the tubing bare, however, if you route it through solid PVC piping, this will limit pressure on or damage to the soft flexible tubing. A stopcock or valve is attached to the end of the tubing. During sampling, you can use a small electric or hand-pump to apply suction (if you have a medium to large cup), but for small samplers, it is simple to attach a syringe and withdraw the plunger. Place a dowel or rod to keep the plunger extended. Over several hours, the syringe will extract pore water from the vicinity of the porous cup. It may take 24–48 h in some instances. If using small samplers, it is advisable to install several at each target depth or horizon. You can then bulk the extracted water to ensure there is sufficient sample for analysis. It is not uncommon for some samplers not to work; they are delicate, and installation should be done with care. Simply striking a pebble can damage the

Fig. 6.26 This researcher is preparing pore water samplers by applying a slurry of soil (kaolinite can also be used) and water sampler before insertion into the pilot hole. This ensures good contact between the probe and the soil. *Source:* Sara Vero.

connection between the porous cup and the flexible tubing. When you begin your sampling campaign you can draw through a volume of water and discard whatever was retained within the tubing. Avoid touching the porous cup with your bare hands. Oils on your skin can block the pores, so always wear laboratory gloves when preparing and installing. When backfilling the soil pit (if you are using a pit-face approach), you should maintain the horizons as much as possible. In other words, try to return soils to the horizons from which they were excavated. This will help maintain the integrity of the profile.

A common question is from what radius is the sample obtained. Unfortunately, this is hard to define and will be influenced by the diameter of the sampler, size of the pores, texture of the soil, and to what degree it is saturated. You may need to accept a degree of uncertainty around this, but realistically, most samples will only be obtained from an area of a few centimeters. Pore-water samplers do seem to particularly attract interference from small wild animals, particularly rabbits and rodents. This is another good reason for accessing tubing via a sealable access port.

References

Baker, D.B., Johnson, L.T., Confesor, R.B. and Crumrine, J.P. (2017). Vertical stratification of soil phosphorus as a concern for dissolved phosphorus runoff in the Lake Erie basin. *Journal of Environmental Quality* 46(6), 1287–1295.

Crozier, C.R., Naderman, G.C., Tucker, M.R. and Sugg,R.E. (1999). Nutrient and pH stratification with conventional and no-till management. *Communications in Soil and Plant Analysis* 10(1&2), 65–74.

Delgado, J.A., Shaffer, M., Hu, C., Lavado, R.S., Wong, J.C., Joosse, P., Li, X., Rimski-Korsakov, H., Follett, R., Colon, W., Sotomayor, D. (2006) A decade of change in nutrient management: a new nitrogen index. *Journal of Soil and Water Conservation* 61(2), 62A–71A.

Logsden, S., Clay, D., Moore, D. and Tsegaye, T., (eds). (2008). *Soil Science: Step-by-Step Field Analysis.* Madison, WI: SSSA.

Sharpley, A.N., Weld, J.L., Beegle, D.B., Kleinman, P.J.A., Gburek, W.J., Moore, Jr., P.A. and Mullins, G. (2003). Development of phosphorus indices for nutrient management planning strategies in the United States. *Journal of Soil and Water Conservation* 58(3), 137–152.

Soil Science Division Staff. (2017). Soil survey manual. In C. Ditzler, K. Scheffe, and H.C. Monger, (eds). USDA Handbook 18. Washington, D.C.: Government Printing Office.

Soil Survey Staff. (2014). Soil survey field and laboratory methods manual. In R. Burt, and Soil Survey Staff, (eds). Soil Survey Investigations Report No. 51, Version 2.0. Washington, D.C.: U.S. Department of Agriculture, Natural Resources Conservation Service.

Sun, S.Q., Cai, H.Y., Chang, S.X. and Bhatti, J.S. (2015). Sample storage-induced changes in the quantity and quality of soil labile organic carbon. *Scientific Reports* 5, 17496. https://doi.org/https://doi.org/10.1038/srep17496

Thien, S.J. (1979). A flow-diagram for teaching texture by feel analysis. *Journal of Agricultural Education* 8, 54–55.

7

Water Techniques

When considering an experiment involving water sampling, it is important to recognize that the hydrologic system consists of several interacting components. Unlike soil, that changes comparatively slowly, water is subject to dynamic fluctuations in levels, chemical composition, and biophysical parameters. For freshwaters, the characteristics measured at a particular point represent the influence of water draining to and from that location; both over land and through the soil and groundwater. The area from which the water is drawn is referred to as the "catchment" or "watershed". Within that watershed are surface waters (Fig. 7.1) (e.g., rivers, lakes, and streams) and two regions of sub-surface water: groundwater held in the permanently saturated rock below a water table, and pore water held in the soil and variably saturated rock above the water table. This variably saturated region may be known as the "unsaturated" or "vadose" zone. Since water moves throughout these interrelated regions, you

Fig. 7.1 This researcher is sampling lake water via an access hole drilled in the ice. *Source:* Derek Gibson.

Fieldwork Ready: An Introductory Guide to Field Research for Agriculture, Environment, and Soil Scientists,
First Edition. Sara E. Vero.
© 2021 American Society of Agronomy, Inc., Crop Science Society of America, Inc., and Soil Science Society of America, Inc. Published 2021 by John Wiley & Sons, Inc.
doi:10.2134/fieldwork.c7

should understand that a sample obtained at any particular point will be influenced by past interactions with the pathways through which it has traveled. There are many excellent textbooks (e.g., Bevan et al., 2010; Winter et al., 1998) that provide detailed explanations of these concepts. It is worthwhile to spend time becoming familiar with the principles of watershed and catchment hydrology before embarking on any sampling regime.

Surface Water Sampling – Freshwater

The chemical composition, ecologic diversity, temperature, pH, and clarity of surface water can be used to indicate its quality. Water quality standards are typically defined by thresholds prescribed in legislation (e.g., the E.U. Water Framework Directive or the United States Clean Water Act), and many nations operate ongoing monitoring to evaluate trends in chemical concentrations and biological quality (as determined by the presence or absence of key indicator species). Typically, nutrients such as phosphorus or nitrogen that may facilitate eutrophication (the excessive growth of plants or algae) are measured as standard. Other contaminants that may pose a threat to human or ecological health may also be considered in some scenarios and locations. Such contaminants include (but are not limited to) herbicides such as diuron or atrazine (among many others), pharmaceuticals such as acetaminophen, heavy metals, and emerging contaminants such as microplastics, caffeine, and even ketamine. Water sampling may also be conducted to evaluate the effects of terrestrial processes such as farming, wildfires, urbanization, etc., and to provide context for ecological research into plant and animal species living in or near a watercourse (Fig. 7.2).

When accessing a watercourse take care not to collapse banks, disturb the bed sediment, or inadvertently alter the channel. Some freshwater species such as the pearl mussel or spawning fish are highly vulnerable to disruption and must be protected. Do not engage in any sampling that threatens a protected species, and if possible, minimize stream disruption during ecologically sensitive periods such as the breeding season of native species. You may wish to consult a local ecologist when planning instream sampling or experiments, as they will be familiar with the species in your area.

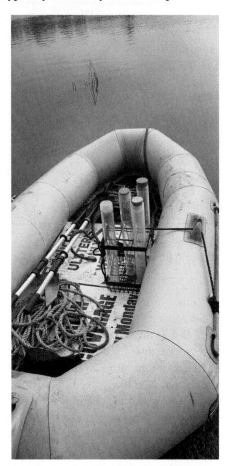

Fig. 7.2 Sampling watercourses may require access by boat. This should be considered when planning your experiment. You may need specialist training if you are not familiar with their operation. *Source:* Julie Campbell.

Instream Parameters

Stream Width and Depth

Stream width and depth are particularly informative if you are returning to the same sampling locations on multiple occasions, as changes may occur seasonally, at different flows or due to structural changes to the channel. For relatively narrow streams, you can simply use a tape to measure the width of the flowing water (Fig. 7.3). Keep in mind, this is not the same as the width of the channel (i.e., relatively static) but rather should be measured at the water surface.

Fig. 7.3 Accurately recording stream width and cross sectional area is important for stream gauging and estimating flow. Details of flow gauging are available in USGS (2006). *Source:* Brandon Forsythe.

You may also be able to measure depth using a tape or measuring pole. A staff gauge is used for measuring water level at a fixed location in a river, stream, or lake (Fig. 7.4). This device is simply a tall measuring staff that is secured in a fixed position to the bank. The zero marker corresponds to a fixed datum, usually the streambed. Surveying equipment can be used to determine the elevation of this datum relative to sea level. The staff gauge should be located in an area that is unlikely to be physically altered or interfered with and not likely to suffer from erosion of the bed sediment or from bank collapse as these could compromise the reading. In other words, if the base of the stream erodes away, then the zero measurement will no longer be accurate, and the real water level will actually be deeper than suggested by simply reading the gauge. Water level is easily read from the staff gauge; you simply note what height the surface equates to. This provides a snapshot measurement; however, high temporal resolution measurements can also be taken using a pressure sensor and a stilling well. A stilling well is essentially a PVC pipe that is affixed in the stream, slightly above the streambed. The pipe should be perforated (this can be done by drilling holes) to allow water to move completely freely in and out and so the water level within the pipe will equate to that in the stream.

The pressure sensor should be suspended inside the pipe using a non-stretch cable (coated steel is ideal) at a depth that is constantly below water but not resting on the streambed. The pressure sensor will detect differences in pressure at that depth, which is proportional to the height of water above. There are several different types of pressure sensors; however, their basic principle consists of a deformable component (such as a diaphragm) known as a "force-summing device" and an electrical transduction element. The force-summing device is distorted by the pressure (i.e., the overlying water) applied to it. The transducer translates this into a pressure measurement by detecting changes in voltage, capacitance, or resistance in the circuit as a result of the deformation. Detailed description of pressure transducers is provided in Freeman et al. (2004).

It is important that the cross-section of the watercourse at the gauging location is fixed to accurately calculate discharge. Fixed structures such as weirs or flumes are often constructed to ensure a consistent cross-section at gauging and sampling locations along the watercourse.

Fig. 7.4 A staff gauge is a measuring rod fixed to the river bank. This can be used to read the height of the water above datum. *Source:* Sara Vero.

Wiers are barriers (or small dams) that partially obstruct flow at a gauging point. Water is thus forced through a channel of known dimensions. Since flow is partially obstructed, the depth of water behind the weir will increase. The difference between the height of the water behind the weir (as measured with a gauge or pressure transducer) and the crest of the weir indicates the head (H). Head is then used to calculate the discharge (Q) based on the dimensions of the channel. Formulas for these calculations vary depending on the design of the weir. Commercially available weirs are typically supplied with calibrated look-up tables or software. Weirs are relatively inexpensive to purchase or construct and can be adapted to site-specific conditions. They are also often suitable for measuring low flow rates, particularly V-notch weirs, which have a narrow opening.

Flumes are structures that create a narrowed channel of a well-defined shape (Fig. 7.5). Various configurations are available, including H-type, trapezoidal and Parshall (and others). Selection of configuration depends on the cross-section of the existing channel. Flumes require little drop in height compared with weirs, so are suitable for measuring flow in streams with a very flat gradient. There is no obstruction in flow, as with weirs. Installation costs of flumes can be greater than weirs; however, they are capable of measuring higher flow rates. Both weirs and flumes require regular maintenance as collection of sediment or debris can obstruct flow, resulting in inaccurate measurements. When selecting weirs or flumes, you should consider the cost of installation, rates of flow you anticipate, and the level of maintenance required.

Fig. 7.5 Flumes like these have a fixed cross-sectional area which allows accurate recording of water flowing through. *Source:* Brandon Forsythe.

Physico-Chemical Parameters

Certain parameters are liable to change or deteriorate during transport of samples to the laboratory. For example, measurements of temperature or electrical conductivity in the laboratory are likely to be different from when that sample was taken from a stream. Many commercially available handheld probes incorporate a number of sensors and are capable of measuring multiple parameters at once (typically referred to as "multiparameter probes"). It is common to encounter sensors for measurement of temperature, dissolved oxygen (DO), pH, electrical conductivity (EC), and salinity incorporated within a single handheld unit. Multiparameter probes can be calibrated in the laboratory and are generally very simple to use in the field. Many have a memory function that you can use to record readings; however, it is recommended to also fill in field sheets. When using any sensors or other equipment in the field you should record the make and model of the device on your field sheet or notes. Research groups often have several units of the same sensor or device. If this is the case, you should record which one you use. This will help when interpreting the results later on and is important from a quality control perspective. If you are taking a physical water sample (a "grab sample"), it is advised that you measure the instream parameters at the same time. These can help with analysis and interpretation of your results.

Using a Multiparameter Probe

Using a multiparameter probe is relatively straightforward. Prior to fieldwork you should calibrate the sensors in the laboratory. This will vary depending on the make and model, so consult the manufacturer's instructions. Calibration should also be done at regular intervals.

In the field, the multiparameter probe should be fully submerged below the water surface, ideally around the middle of the water column. It may take a few minutes for the reading to stabilize. Many devices will emit a "beep" to tell you that the reading has stabilized. After this, you can save the measurement using the memory function and also record it on your field sheet. Avoid turbulence if possible as this will bring air bubbles in contact with the measurement cell and it may take longer for the reading to stabilize (particularly DO readings). If the stream is very shallow and it is not possible to submerge the probe, you may collect a sample in a clean sample bottle and submerge the probe. You should record whether the probe was submerged in stream or in a sample bottle (Fig. 7.6).

Fig. 7.6 Multiparameter probes can also be lowered down monitoring wells to measure the temperature, conductivity, pH or other parameters of groundwater. Use a secure cable to prevent losing your equipment down the well! *Source:* Brandon Forsythe.

It may be helpful to understand how the various sensor types work. Let us discuss some of the parameters in more detail.

Temperature – Surface water temperature can vary with position along the watercourse, depth in water column, shading by overhanging vegetation, contribution of groundwater, runoff, snow-melt, and other factors. Water temperature is important because it influences the life cycles of fish, diatoms, and invertebrates. Crucially, elevated temperatures can facilitate algal prolifera-tion where nutrients are not limiting. Typically, changes in water temperature are slow to respond to changes in air temperature, particularly where the watercourse in groundwater-fed. There is a wide variety of controls on stream temperature, not limited to urban discharges, vari-ations in groundwater contribution, and snowmelt runoff. A thorough review is provided in Webb et al. (2008). In-stream temperature can be measured using an electrical probe or a ther-mometer. Measurements may be reported in Celsius (°C) or Fahrenheit (°F) as per local conventions.

Dissolved oxygen – Dissolved oxygen (DO) is the amount of gaseous O_2 in solution. This can *only* be measured in-stream as it will be immediately altered by grab sampling. Dissolved oxygen is influenced by temperature, turbulence, aeration, and by respiration of aquatic plants and animals. Water can contain greater DO at cool temperatures relative to warmer temperatures. Eutrophication may cause low DO, which can impair invertebrate and fish species. Dissolved oxygen can be reported in absolute values ($mg\,L^{-1}$) or as a percentage satu-ration (in other words, as a percentage of the maximum DO which water can hold at that temperature). Measurement of DO is typically done by either of two approaches: optical sen-sors or electrochemical sensors. Optical sensors work by emitting a light of known wave-length and then measuring either the intensity or the lifetime of illumination. These are inversely related to oxygen content. In other words, the intensity and lifetime of illumination are greatest when no oxygen is present and decrease when greater amounts of dissolved oxygen is present. Accuracy of these sensors is influenced by water temperature so many probes incorporate a thermometer and a mathematical correction factor in the software. Electrochemical sensors work by measuring the electrical current passing between a cathode (electron donor) and an anode (electron acceptor), which are enclosed within a permeable membrane through which O_2 diffuses. Electrochemical sensors are influenced by barometric pressure, temperature, and salinity. There is generally strong agreement in results between the two approaches; however, the electrochemical type has a quicker response time and so may be more convenient if a large number of samples are to be taken.

pH – The pH indicates the acidity or alkalinity of a sample (1 (highly acidic) to 14 (highly alka-line), 7 is neutral). The pH scale is logarithmic, meaning that each value is ten times greater or lesser than the previous value, depending on direction. In other words, a pH value of 9 is 10 times more alkaline than a value of 8, while a value of 3 is 10 times more acidic than a value of 4. The pH of a waterbody influences the buffering and mobilization of nutrients within the sedi-ment. It also effects the suitability of the watercourse as a habitat for microbial, invertebrate, vertebrate, and plant species, each of which have their own optimal ranges. Sensors for measur-ing pH work by relating difference in ion activity between the sample solution and a reference within a glass electrode. You can also measure pH of grab samples in the laboratory or use test strips.

Electrical conductivity – Electrical conductivity (EC) is the capacity of water to conduct an electrical current. This capacity is influenced by the presence of dissolved ions (such as salts) and by temperature. High EC values are often associated with salinity (Rusydi, 2017). Sudden or isolated changes in EC observed along a watercourse can be indicative of the presence of a point source of pollution such as a wastewater treatment facility, farmyard or urban runoff. Sensors that measure EC work by detecting the resistance of an electrical current moving between electrodes that are set at a fixed distance apart. MilliSiemens per cm ($mS\,cm^{-1}$) are the unit of measurement.

Salinity – Salinity refers to the concentration of dissolved salts and indicates the "freshness" of the water. Highly saline surface waters (excepting marine water) are said to be "brackish." High salinity may result from depletion of contributing groundwaters, industrial or mining discharges, or intrusion of marine waters. As the salinity of freshwaters increase, their ability to support diverse freshwater biota decreases as many species cannot tolerate such conditions (Nielsen et al., 2003). Salinity can be reported in either parts per mille (‰) or mg L ($mg\,L^{-1}$). Freshwater generally exhibits $<3000\,mg\,L^{-1}$, while salt water $>35,000\,mg\,L^{-1}$.

Sampling for Laboratory Analysis

Fig. 7.7 When grab sampling, rinse the bottle with water from the watercourse before taking the sample. *Source:* Sara Vero.

Samples may be returned to the laboratory for chemical or microbial analysis (Fig. 7.7). Consult your laboratory standard operating procedures (SOPs) for preferred methods. The volumes need will depend on the nature of the analysis to be conducted and on the standard procedures in that laboratory. Clean, sterile, labeled bottles, or containers should be used for sample storage. Plastic bottles are safer from a breakage perspective; however, glass containers may be used, especially for microbial analyses. Glass containers can be sterilized and pose little risk of leaching or degradation. Whichever material is preferred, secure, screw-top lids should always be used, and bottles should be stored upright during transit to prevent spills and leaks. Grab samples can be taken directly from the watercourse as follows:

1) Ensure that you are securely positioned and not at risk of slipping into the watercourse.
2) Measure and record in-stream parameters using a handheld probe. Avoid disturbing the bed or bank sediment.
3) Rinse the sample bottle with water from the watercourse.
4) Fill the sample bottle to the required level and seal securely.
5) Record who took the sample, the exact location, date, time, temperature, and any observations of the site such as clarity of the water, presence of point sources etc.

Since water is flowing, observations at a specific moment may or may not be wholly reflective of the in-stream conditions at other times. Date and time are therefore crucial as they allow your results to be interpreted with respect to the activities taking place in the catchment at that time and the discharge of the watercourse.

For many small and shallow streams measuring only a meter or two across, a single sample may be sufficiently representative of the water flowing through the channel. However, for wider rivers, the composition may vary across the cross-section. In these instances, the equal width increment (EWI) method can be used to collect a composite sample. To do so, you divide the width of the channel into equal sections and take a proportional sample from each segment. These samples are then integrated. Detailed and statistical recommendations for EWI sampling are provided by Edwards and Glysson (1999).

If the water is deep, difficult to access or has high banks you can use a sampling pole or a dipper to lower the sample container from a safe location. Dippers can also be used to obtain samples from a particular target depth (Figs. 7.8 and 7.9). You should always record the depth from which

Fig. 7.8 Multi-parameter probes can be lowered deeper within the water-column using cable. Make sure to record the depth at which you are measuring. *Source:* Derek Gibson.

Fig. 7.9 Extendable poles ('dippers') can be used to collect a sample from deeper within the watercourse, or to sample from a bridge or bank. *Source:* Katie O'Reilly.

samples are obtained. This is particularly important when sampling still water such as lakes or reservoirs. A challenge in this is preventing contamination or mixing of the sample when bringing it to the surface. A bailer can be used for this; these are chambers that can be lowered to desired depths using a cable and can then be triggered to close before returning above water. The target depth will depend on the objectives of your study and the overall depth of the watercourse. When sampling lakes, it is common to take several discrete samples from various depths as stratification can occur within the water column. You may also take this approach in deep rivers. Alternatively, a composite sample can be taken, consisting of subsamples from several depths. For shallow streams, it is common to take a single sample from the middle of the water column. Consult the literature related to your research objective and consider the morphology of the watercourse in question.

Grab samples should be transported to the laboratory or to cold storage at the earliest opportunity (Fig. 7.10). This is because microbial and biochemical processes will influence the chemical composition of the sample, so analyses should be conducted promptly to reflect in-stream values. A cool-box with ice packs should be used to preserve samples in transit. Certain parameters such as total oxidized nitrogen or soluble reactive phosphorus are measured on filtered samples. This should be done as soon as possible after taking the sample. You may bring paper filters and suitable containers to do this in the field, or you can use syringes with attachable filters to speed up the process. Remember to label both the filtered and unfiltered samples. Some parameters have short "hold times." This refers to the amount of time that the sample may be stored for prior to analysis without impairing the quality of the analysis. You should consult your lab SOPs in relation to this, but some general rules are shown in Table 7.1.

With all water sampling, bear in mind the weight of the sample. Each liter of water weighs a kilogram. You may need large volumes or numbers of samples. Be realistic about how much sample you can safely carry and always adhere to good manual handling protocols.

Fig. 7.10 Water samples should be stored in cool-boxes or on ice. This researcher is carefully documenting the details of these samples on corresponding field sheets. *Source:* Nikki Roach.

Table 7.1 General hold times and storage for common parameters.

Parameter	Storage	Hold time
Phosphorus (total, particulate & dissolved)	Cool (4 °C)	<24 h
Nitrogen (total, dissolved organic N, nitrate, ammonia)	Cool (4 °C)	<48 h
	Freezing (−20 °C)	30 d
Dissolved organic carbon	Cool (4 °C)	<48 h
Metals	Cool (4 °C)	6 mo
Biochemical oxygen demand	Cool (4 °C)	<48 h
Bacterial analysis	Cool (4 °C)	6–24 h

Autosamplers

Autosamplers are mechanical devices programmed to siphon a water sample from a waterbody using a pump (Fig. 7.11). This can be done at scheduled intervals, in response to stream conditions monitored using an instream sensor and datalogger or by remote trigger. Samples are collected in prelabeled bottles and must be retrieved and returned to the laboratory for analysis. There are major logistical advantages to using autosamplers. They allow sampling during times or conditions, which would be challenging or uncomfortable for a human technician (e.g., overnight or throughout storms) and can be equipped with refrigeration to maintain the quality of samples until they can be retrieved. It is important to remember that the hold times for your desired analysis still apply to samples obtained using an autosampler, and you should consider both the time

Fig. 7.11 An autosampler consists of a pump, datalogger and sampling bottles, shown here. In this design, a mechanical arm rotates either on a timer or based on flow if the sampler is connected to a flow-meter and deposits an individual sample in each bottle. *Source:* Sara Vero.

spent in transit and the maximum duration that the sample was contained in the autosampler. Different brands and proprietary designs are available but, in most instances, they consist of a pump and a programmable timer or modem to initiate sampling. A carousel is loaded with sample containers (often up to 24 depending on their volume and on the size of the autosampler). This carousel rotates allowing each container to be filled in sequence.

Samplers should be placed on a level, stable bank above the maximum height that the river can reach and secured to the ground by bolts or a tether. If you have the budget and resources, you can house them within fiberglass shells known as kiosks. Although autosamplers are weatherproof, kiosks help protect them from interference by animals or third parties. Autosamplers should be installed at stream cross-sections, which are relatively uniform (Fig. 7.12).

Inlet tubing should be routed from the sampler to the stream. The stream-end of the tube should be open in the direction of flow, at the mid-depth of the water column. The pipe should be adjusted so that it is always fully submerged. Positioning the pipe too low can allow sediment from the bed to be sucked into the sampler. You should also locate the opening away from banks for the same reason and avoid areas of turbulent flow. A steel stake should be driven into the streambed and the pipe opening affixed to it. Using a clamp with a screw will allow the pipe to be moved up and down the stake as required, however, cable ties are an inexpensive alternative. The inlet tubing should be as short as reasonably possible (< 25 ft), while maintaining a gradual gradient from inlet to autosampler. Make sure there are no bends or pinches in the tubing which could restrict flow, and remember, high-flows and storms can bring debris which may damage the installation, so it is critical to inspect regularly and make additional visits after heavy rainfall or winds. It is common to capture discrete samples using autosamplers, after which the carousel rotates allowing a new sample to be delivered into the following container. Alternatively, flow-weighted composite sampling (FWCS) can also be achieved with autosamplers. In this approach, real-time discharge data regulates the timing and quantity of sample extracted from the stream. These individual samples are composited to provide a flow-weighted mean concentration, which can then be divided by total discharge to give the load (Cassidy et al., 2018). This approach is most appropriate for relatively conservative

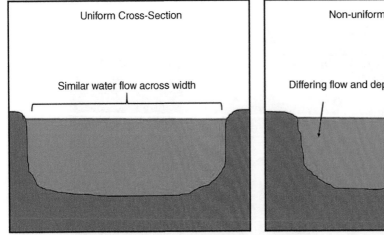
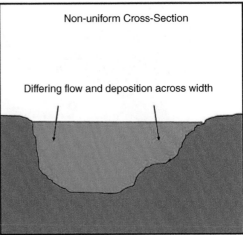

Fig. 7.12 Autosamplers should be installed at uniform cross-sections. Where sampling is required at non-uniform cross-sections, multiple grab samples may be required across the channel. *Source:* Sara Vero.

parameters (e.g., total phosphorus or total nitrogen) and less so for those fractions which may be biologically or otherwise altered during storage (e.g., reactive phosphorus). A major advantage of this approach is that it lowers the influence of temporal dynamics observed in discrete sampling approaches which can lead to either over- or underestimation of total loads. A composite approach overcomes this and furthermore, reduces the labor required by high-temporal resolution discrete sampling. This might be crucial where resources are limited, traveling to study sites is particularly arduous or when multiple locations are being monitored. You should consider the trade-off between resolution and practicality with respect to (i) the particular parameter of interest and (ii) the specific objective of your project. Cassidy et al. (2018) suggested that a combination of FWCS and grab sampling approaches could be considered.

It is common to also install flow-gauges or probes adjacent to autosampling locations. As discussed earlier, flow gauges (which may involve fixed structures; flumes or weirs) allow discharge to be monitored (Fig. 7.13). When coupled with concentration data obtained via sampling and analysis, this allows loads to be calculated. Crucially, frequent sample collection and measurement of discharge at high temporal resolution allows more accurate estimation of total nutrient loads than less frequent measurements. The latter approach may lead to over- or underestimation of nutrient export and may miss relatively short peaks or troughs. In-stream probes should be similarly affixed at appropriate depths and wired to their dataloggers. Dataloggers are discussed in the environmental monitoring section and the same principles apply here. This approach allows the relationships between in-stream parameters such as temperature, discharge, or oxygen and parameters measured from the physical sample (such as nutrient concentration).

Fig. 7.13 This open channel velocimeter sits on the streambed and measures flow based on the Doppler principle. *Source:* Sara Vero.

Sediment (Bed, Bank, and Water Column)

The bed and banks of waterbodies contain a wealth of information as they reflect the legacy of physical processes such as erosion and deposition, and of biological processes such as nutrient cycling. The physical, chemical, and organic constituents of sediment transported through a river cross-section are also an important indicator of erosional processes in the contributing catchment. As such, there is no strict criteria for the location of sediment sampling (Radtke, 2005; Murdock and McKnight, 1994). It is crucial when sediment sampling to disturb the streambed as little as possible. Enter the watercourse with caution and always position yourself downstream of the sample so that any disturbance you cause is not captured in your sample.

For very shallow streams or ditches (Fig. 7.14) (those which can be safely waded by the technician), bed samples can be collected simply using a trowel and stored in secure labeled bags (Shore et al., 2015). You should record the depth from which your samples are taken. If the bed is relatively uniform, you should take several samples from the cross-section and composite. You may need to divide nonuniform beds into separate sections and sample individually. Sampling banks can be conducted similarly. Using a trowel, excavate samples from your required depth and location and label accordingly. Alternatively, a cylinder of known volume can be pushed to a measured depth into the sediment. The sample within the cylinder is then extracted and homogenized. This allows a more representative sample to be taken and limits the loss of smaller sediments. For nonuniform or wide cross-sections, or where differing patterns of deposition and erosion occur (e.g., at bends), it may be preferable to analyze each grab sample discretely.

Corers allow you to sample deeper sediment (Fig. 7.15). As sediment is accumulated gradually over time, these deeper analyses can provide information on long-term trends. The procedure is similar to that for soil augering. Place the base of the corer on the stream bed and push directly downward into the sediment. This may not be possible in very stony beds. Once you have reached the target depth or when the corer cannot be inserted any further, withdraw upward (Fig. 7.16). Some corers apply a sleeve or liner to the sediment column. This can prevent fine sediment from

Fig. 7.14 These researchers are using augers to sample submerged soil/sediment in a shallow wetland. *Source:* Jaclyn Fiola.

Fig. 7.15 Sediment in deeper river or lake beds can be sampled using a variety of corers, or augers in this picture. The researchers here are using extension rods to push the sampler deeper into the bed. *Source:* Jaclyn Fiola.

Fig. 7.16 Sediment cores can be several metres long, depending on the nature of the bed and the length of the sampler. This can be difficult to manuevre so you may need assistance from your team-mates. *Source:* Derek Gibson.

being lost during extraction and transport. Freeze-coring can be used for very wet or unconsolidated sediments likely to be lost during extraction. This method involves first driving a double-walled cylinder or container into the sediment. The interior of the walls is then filled with a coolant such as ethanol or dry-ice and the entire chamber is extracted once the sediment has frozen. Individual slices of the core can then be separated. This method is particularly useful in highly stratified sediments (e.g. in lakes). Freeze-corers are available from various suppliers, although several designs for their construction are also available (e.g., Franchini and Zeyer, 2012; Trost et al., 2018; Schimmelmann et al., 2018; and others). As with any time you construct equipment, always test your corer thoroughly before deploying to the field.

Once you have extracted your sediment core, you should cap both ends of the cylinder and label (Figs. 7.17 and 7.18). Cores should be carefully laid in boxes or tubes such as PVC piping. You should analyze layers individually and record the depths from which you take subsamples for analysis (Fig. 7.19). Homogenizing across the length of cores is *not* recommended.

Fig. 7.17 This sediment core has been extracted from the lake bed and is being capped before storage and transport. *Source:* Derek Gibson.

For deep water a grab sampler is required (Fig. 7.20). These are bucket-type devices which are lowered to the bed on a cable and mechanically triggered to close, taking a scoop of bed material. There are several varieties available with different trigger mechanisms and bucket/mouth designs. You should select your grab sampler based on the nature of the bed sediment. Three popular grab samplers are summarized in Table 7.2, but be aware that other designs exist. When sampling, open the mouth of the bucket and lower or winch to the bed. Do so slowly, keeping tension on the cable. Once the bucket has reached the bed and closed, slowly retract it upward. Moving too rapidly can cause fines to be dislodged and lost. Once the sampler reaches the surface, bag and label the collected sample. Be aware that grab samplers won't allow you to disentangle different sediment depths or layers as coring approaches do.

Suspended sediment (SS) are the fine particles (<63 μm) carried in the water column. These particles are eroded from the bed and banks or transported to the watercourse from the contributing upstream catchment via overland flow or laterally through the subsurface. Elevated SS is highly detrimental to the aquatic habitat and may particularly harm sensitive species such as the freshwater pearl mussel. The repealed EU Freshwater Fish Directive (78/659/EEC) proposed a guideline of $25\,mg\,L^{-1}$ suspended sediment (average annual concentration) and the deposition of fine sediment into channel beds can affect the ability of watercourses to attain

Fig. 7.18 In this photo rubber bungs have been used to seal the core and prevent loss of sediment from this lake sample. *Source:* Julie Campbell.

Table 7.2 Grab sampler designs and their descriptions.

Grab sampler design	Description	Suitable for:
Shipek	A heavy, rotating bucket (single jaw) with torsion springs that swings rapidly on hitting the bed	Soft to hard or packed unconsolidated sediments. Not impeded by pebbles <5 cm
Van Veen	Lightweight clamshell bucket (double jaw) with a scissor mechanism that closes automatically on hitting the bed	Top layer of sediment (<20 cm), silt or sand. Not suitable for hard or packed sediment
Petersen	Heavy clamshell bucket (double jaw) with a trip mechanism that closes on hitting the bed and cable slackening	Hard sediments including sand and gravel

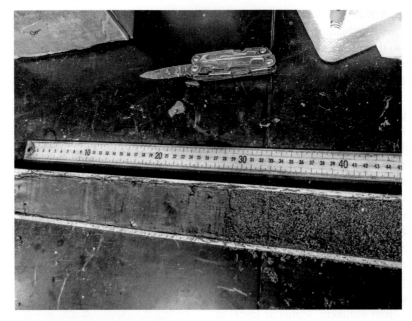

Fig. 7.19 Like soil cores, sediment cores should be measured and sectioned depending on layers. *Source:* Derek Gibson.

Fig. 7.20 This pontoon has been adapted to include solar panels for charging batteries and a winch for collecting sediment cores. *Source:* Derek Gibson.

good ecological status under the EU Water Framework Directive (2000/60/EC). There are several approaches to collection of sediment from the water column itself (Fig. 7.21).

Grab/Autosampling – A volume of water can be collected from the stream either using the standard grab sampling procedure or using autosamplers. Subsequently, these samples are analyzed in the laboratory by the filtration and/or evaporation and thereafter, calculating the mass of solids per volume of solute. Total suspended sediment is therefore expressed in $mg\,L^{-1}$. You may also measure particle size distribution, chemical properties, etc.

Fig. 7.21 The buddy system in action - researchers collecting sediment samples from a wetland. *Source:* Katie O'Reilly.

There are two chief limitations to be aware off. First, similar to all grab sampling, it is a snapshot of that specific moment in time and flow conditions. This may be particularly problematic when assessing sediment dynamics as overland transport is dictated by many factors that change across the year such as crop cover and soil saturation, and by the intensity of precipitation. If a grab sampling approach is selected, you must interpret your data in light of these factors and ideally, should collect samples across a range of conditions.

Turbidity probes – Turbidity is a measure of the optical properties of a liquid. For stream sediment, this indicates the reduction in clarity caused by suspended particles. In very simple terms, turbidity measures how cloudy the water is. Using optical sensors linked to dataloggers, it can be used to provide a high-temporal resolution alternative to direct measurements. Sensors can either be deployed instream (secured within the water column) or ex situ (a water sample is siphoned out of the stream into an instrumented tank). Many multiparameter probes can also be fitted with turbidity sensors. Sherriff et al. (2015) compared instream and ex situ methods and found that the difference between measuring approaches was not significant. The ex situ approach requires maintenance of the pump, tank, and tubing to prevent blockages (particularly after storm events in which large amounts of debris are transported). However, it is less sensitive to spurious readings reflecting disturbance of the bed immediately around the submerged sensor (Lewis and Eads, 2001; Sherriff et al., 2015). Regardless of which approach to instrumentation you use, it is sensible to take grab samples at various flow periods. This allows the relationship between stream discharge and turbidity to be derived for that site.

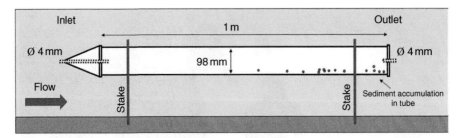

Fig. 7.22 Time-integrated sediment sampler diagram.

Time-integrated sediment samplers (TISS) – This approach (Phillips et al., 2000) utilizes instream collection tubes to accumulate sediment from the water column over a prolonged period (Fig. 7.22). TISS can be constructed using materials readily available at most hardware stores. The typical design of these samplers involves a 1 m long PVC tube, capped at either end, allowing a 4 mm inlet and outlet. A funnel is placed over the inlet to streamline flow. Water flow slows as it passes through the tube allowing sediment to drop out of suspension and be retained until collection. The entire device should be held within the water column by affixing it to posts or a frame within the stream bed. Chains and weights can also be used to suspend the samplers from bridges and other structures.

A TISS collects a bulk sample representative of the total sampling time. Where available, discharge data can aid in the interpretation of these samples including the relationship to storm events. A further utility of these devices is in allowing sufficient samples to be collected for analysis even in low turbidity streams by extending the sampling period (or increasing the number of samplers at one location).

Groundwater Sampling

Monitoring wells are used for groundwater sampling (Fig. 7.23), although springs can also be used to obtain samples from perched water tables in hilly landscapes. Monitoring wells consist of one or more boreholes drilled from the soil surface through the vadose zone into the aquifer (Fig. 7.24), intercepting the water table (Fig. 7.25). The water table will fluctuate over time depending on recharge and abstraction (extraction of groundwater, e.g., for drinking or irrigation). A monitoring well must therefore be deeper than the water table at the time of drilling and ideally, deeper than the anticipated maximum depth of the water table during very dry or drought conditions. Of course, deeper drilling is costlier, so consider (i) the likely depth of the water table within your monitoring period, (ii) the objectives of your experimental design, and (iii) your budget. In most instances, a drilling company will be contracted to establish a monitoring well or transect. They can advise you regarding the best depth and construction based on site specific geology. In many cases, a geophysical survey including ground penetrating radar (GPR) may be conducted to aid in both design and in the interpretation of monitoring results. Ground penetrating radar creates images of the subsurface by emitting high-frequency electromagnetic pulses and detecting the reflected signals. The patterns which are reflected or refracted back to the GPR receiver depends on the characteristics of the soil, rock, voids, or artifacts below. This technology has been widely used in groundwater research to map bedrock and aquifers. Ground penetrating radar devices may be available in your research facility, or you may use external consultants.

Fig. 7.23 This monitoring well has a steel casing above the ground to protect the access point. *Source:* Sara Vero.

Fig. 7.24 Drilling a well. *Source:* Brandon Forsythe.

At its simplest, a borehole is a shaft drilled into the ground that is fitted with a casing (often PVC) to prevent collapse or degeneration over time. Space between the casing and the borehole walls are sealed with bentonite. The seal and the casing also prevent ingress of water from the surrounding area. Only the bottom of the borehole, below the water table, is fitted with a permeable screen and surrounded by porous gravel pack. This ensures that only water from within the aquifer is sampled, and not that which is in transit through the variably saturated soil profile.

If you are fortunate enough to have an artesian well (one which is pressurized due to confinement of the aquifer), you will not need a pump as the water will flow freely.

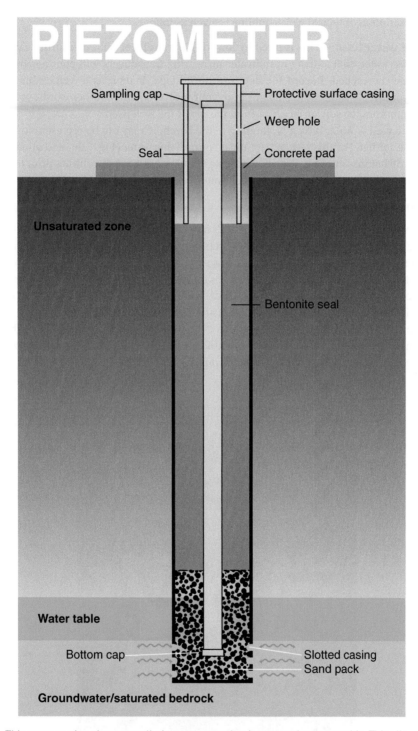

PIEZOMETER

Sampling cap

Protective surface casing

Weep hole

Seal

Concrete pad

Unsaturated zone

Bentonite seal

Water table

Bottom cap

Slotted casing

Sand pack

Groundwater/saturated bedrock

Fig. 7.25 This cross-section shows a well piezometer used to intercept the water table. This allows measurement of groundwater depth and sampling. *Source:* Karen Brey.

A basic description of typical well sampling is given as follows.

1) **Measure water level** – This is done by lowering a transducer on a measuring cable until it reaches the water table. Most groundwater measuring transducers will emit a sound or light when water is reached. Record the depth from the tape. If there is a well-casing above the ground, you should subtract this height from your measurement; in other words, you calculate water depth from the soil surface, not from the top of the well itself.

2) **Purge the well** – Water which is standing in the borehole may not be representative of water within the aquifer. For this reason, you must "purge" this water (Fig. 7.26) and allow the borehole to refill before sampling. Typically purging of at least three well-volumes prior to sampling is recommended. First calculate how much water you need to extract:

$$V = 0.041d^2h$$

Where V is volume in gallons to be purged, d is diameter of well in inches, and h is depth of water in feet (you need to know the depth of the borehole) (USEPA, 2013). You should monitor field parameters such as temperature, pH, ORP, or dissolved oxygen at intervals throughout purging, as their stability will indicate when a representative sample is being drawn from the aquifer. It is typical to purge at least three well volumes; however, you may need to extract more than this if the water remains cloudy or discolored or if the measured parameters do not stabilize.

You can purge the well using a bailer by repeatedly lowering into the borehole and withdrawing. However, this can be laborious and time-consuming. Alternatively, you can use a motorized pump. You can use either a peristaltic or a submersible pump.

Fig. 7.26 Water standing in the borehole should be purged before taking a groundwater sample. *Source:* Brandon Forsythe.

1) If you are using a peristaltic pump, you should run Teflon tubing from the vacuum port of the pump to just above the base of the borehole. Avoid disturbing the base as this can draw silt into the system. Run a second, shorter length of tubing from the discharge port to a bucket or container for collection of the purged water. If using a submersible pump, you simply lower the pump itself 1–2 m into the water column.

2) Turn on the pump. Some pumps will allow you to control the pumping rate; however, others may not specify the rate. If this is the case, you can test the rate by measuring how long in seconds it takes to fill a bucket of known volume.

3) Measure the biophysical parameters (temperature, pH, EC) of purge water at regular intervals. When these parameters stabilize, you may start drawing your sample.

4) Dispose of purged water. Be aware of national or state regulations for disposing of purged water, particularly if it contains potential environmental contaminants. If you are simply pouring it out onto suitable land, make sure that it is sufficiently far away from the drawdown area of the well. Otherwise, that same water is liable to infiltrate back through the soil profile and into the abstraction zone. This will influence the composition of water sampled from that well in future.

3) **Take your groundwater sample** – This should be done as soon as possible after purging the borehole. You may use a Teflon bailer or a pump. If using a pump, you can collect your sample from the discharge pipe into a sterile, labeled flask. Never use the same tubing for collecting your sample as was used for purging the borehole. You should replace this tubing with a new length and dispose of this after sampling. This prevents contamination of the sample. As with surface water sampling, you should rinse the flask or container before filling with your sample. Rates of sampling should be no more than 500 mL min^{-1} for containers greater than 250 mL (USGS, 2006). When purging and sampling a monitoring well, ensure that flow is constant and non-turbulent (Fig. 7.27).

4) **Clean** – Dispose of used tubing. Rinse bailers and other sampling equipment with deionized water before sampling at another location.

Fig. 7.27 When sampling groundwater ensure a steady, non-turbulent flow. These researchers are taking care not to disturb the base of the borehole. *Source:* Brandon Forsythe.

Microbial and Sterile Sampling

Microbial analysis is frequently conducted on soil samples to determine carbon and nitrogen dynamics, gaseous emissions, or contamination with fecal organisms. Samples can be extracted from specific horizons within a soil profile or from across plots or fields. Microbial analysis of water samples can indicate native bacterial communities and provide insight into microbial processing of in-stream nutrients. Microbial analysis is frequently used to determine the potability (suitability for drinking) of that water and as a means to determine the origin of contaminant sources. Fecal indicator bacteria such as *Escherichia coli*, *Salmonella*, or *Enterococci* may cause illness, and large communities can indicate that a waterbody is not suitable for consumption or has been contaminated with potential pathogens. Fecal bacteria are native to the intestines of warm-blooded species, and depending on which bacteria dominate the community, may indicate the host species influencing that waterbody (human or animal). Critically, the length of time a particular species can persist in a waterbody will influence its suitability as an indicator of fecal contamination. If the species dies out very quickly, it may be difficult to capture in a low-frequency sampling regime. Conversely, if the species establishes a community in the watercourse itself, it may be difficult to differentiate between relatively recent inputs from native bacteria.

Regardless of what substrate you are sampling (soil, water, silage, manure, etc.), all equipment used for microbial sampling must be completely sterile prior to deploying to the field. This can involve autoclaving, pressure sterilization, acid washing, or other methods. You should consult laboratory SOPs to determine recommended methods. Once you have sterilized equipment, it must be stored in sterile containers and handled only when wearing clean gloves. It is very easy to recontaminate equipment! Also, if acids have been used, your equipment must be completely rinsed as trace amounts can kill bacteria.

Samples for microbial analysis have relatively short hold times. These may vary depending on the purpose of your sampling (i.e., testing for potability vs. testing for source tracing). As a rule of thumb, samples should be processed within 24 h of sampling. Certain bacteria such as *Clostridium perfringens* may have shorter hold times. You should review the literature and SOPs in relation to your purpose and species of interest when planning your sampling campaign. Due to the time constraints, it is generally best to collect these samples last, immediately before leaving the site. Be aware of what time your laboratory will accept final samples; it would be a waste of your hard work if you arrive too late for analysis to be initiated.

If you have several measurements or activities to do in a single day or fieldtrip, it is generally best to conduct microbial sampling last. This helps adhere to the correct hold times and also, limits the opportunity to contaminate samples by alternating between sterile and non-sterile equipment. Before beginning sterile sampling, you should disinfect your hands and equipment thoroughly. If you have a teammate with you, microbial sampling is an ideal time to implement the Clean Hands Dirty Hands protocol with one partner taking the sample and the other recording site details and handling nonsterile equipment. Basic approaches for soil and water sampling are described here. However, as with all sampling, you should consult the specific protocols used in your laboratory.

Soil and Sediments

1) Wear clean double gloves to prevent contamination of the sample.
2) Sterilize sampling equipment (e.g., augers, corers, or trowels) by wiping down with 70% EtOH.
3) Extract your soil sample using your auger, corer, or trowel. If sampling across a field or plot you may wish to collate samples by mixing in a clean plastic bag and then subsampling replicates. Avoid stones or root matter.
4) Store the collected samples in pre-labeled sterile tubes. Seal tubes immediately.
5) If you are very close to your laboratory and will be able to store within an hour, you can transport samples in a cool box with icepacks. In warm conditions, this may not be a suitable method of transport. Alternatively, you can flash-freeze the samples in the field using liquid nitrogen. Samples in their containers are simply placed within a dewar or cryo-can of liquid nitrogen. It is essential that safety protocols are followed as liquid nitrogen can severely burn skin instantly. Never inhale the vapor. Wear a face shield, apron, and cryo-gloves. This can be awkward in the field, so it is helpful to have a teammate with you. Always carry the liquid nitrogen container upright. Labels very often fall off of sample containers during storage in liquid nitrogen as it degrades the adhesive. You may use permanent marker to write directly on sample bottles or fasten tags using rubber bands.

Water

1) Wear clean double gloves to prevent contamination of the sample.
2) Position yourself downstream of the sample so as not to contaminate the water flowing into the bottle. Do not disturb the bed or bank sediment.
3) Only remove the bottle cap immediately before submerging in the waterbody or open the cap below the surface. Fill the bottle by pointing into the current. You may use depth or width increments depending on the channel dimensions, or take a single point sample, as appropriate.
4) Cap the bottle tightly, allowing 2.5–5 cm headspace.
5) Bottles should be stored upright in a cool-box, along with icepacks or gel coolant bags. You can also purchase ice at a grocery store or service station. Generally, no greater than 1–4 °C is suitable.

Sample Preservation

After taking any sample, it is necessary to avoid or limit any changes occurring prior to analysis. This means that your analysis indicates whatever the real physical, chemical, or biological conditions were in the field. It is rarely possible to achieve this perfectly, as the act of sampling itself often influences the sample, even without contamination or degradation. Degradation may include physical processes such as adsorption or release, or biological processes such as microbial respiration. For water samples, the main methods of preservation are chilling, freezing, and chemical preservation via acidification.

Filtration
Filtration is often conducted either in the field or after returning to the laboratory to remove large particulate matter and some algae and bacteria. Filtration simply involves separation of liquid and solid particles using a pores filter or membrane. The size of pores will dictate which particulates are retained. Pore size is measured in micrometers (μm). A range of filter sizes are available, although 0.45 μm is particularly common for differentiating between total and dissolved fractions of phosphorus, nitrogen, metals, etc. Notably, 0.2 μm is recommended for dissolved organic carbon (DOC). Syringe filters are particularly useful in the field. Alternatively, paper filters and funnels can be used, although this is typically done in the lab immediately after returning from the field. Filtration can be slow, particularly if the sample has a large amount of particulate matter, so schedule accordingly.

Chilling and Freezing
Chilling (storage at <4 °C) and freezing (storage at <0 °C) limit chemical and biological reactions after sampling. In both methods, you must allow headspace (empty space in the sample container) to allow for any expansion of the fluid during storage; otherwise the containers can burst. All samples should be labeled clearly prior to storage in freezers of refrigerators. Frozen samples should be thawed slowly until they are completely liquid and should be mixed thoroughly prior to analysis. Similarly, chilled samples should be allowed to return to ambient temperature prior to analysis.

There are a couple of safety considerations in relation to freezing and chilling. It is common for cold rooms and refrigerators to have temperature alarms that alert you if the target temperature is not reached. They require regular servicing and quality assurance checks. Consult your laboratory records to see that they are up-to-date prior to sampling. Walk-in cold rooms must have internal safety handles to prevent accidental lock in. Always keep your cold rooms and refrigerators tidy, and dispose of any old or expired samples appropriately.

Acidification
Acidification is essential for analysis of trace metals in water samples. This requires lowering the pH of the sample to less than 2. At low pH, the precipitation of iron (Fe), copper (Cu), nickel (Ni), aluminum (Al), and zinc (Zi) is minimized. Acidification can be achieved by addition of nitric acid (HNO_3). This is typically done on returning to the laboratory and after the samples have been filtered. Acidification should be done gradually, using a pH probe to ascertain when the target acidity has been reached. Acidified samples should then be chilled (<4 °C) for up to 6 mo. Sulfuric acid (H_2SO_4) can also be used to inhibit bacterial cycling of nitrogen for up to 7 d in chilled conditions.

References

Beven, K.J. (2010). *Rainfall-Runoff Modeling: The Primer*. New York: Wiley.

Cassidy, R.C., Jordan, P., Bechmann, M., Kronvang, B., Kyllamar, K. and Shore, M. (2018). Assessments of composite and discrete sampling approaches for water quality monitoring. *Water Resources Management* 32, 3103–3118. doi:https://doi.org/10.1007/s11269-018-1978-5

Edwards, T.K. and Glysson, G.D. (1999). Field methods for measurement of fluvial sediment. In: USGS, (ed). Techniques of Water Resources Investigations of the U.S. Geological Survey. Washington, D.C.: United States Geological Survey.

Franchini, A.G. and Zeyer, J. (2012). Freeze-coring method for characterization of microbial community structure and function in wetland soils at high spatial resolution. *Applied and Environmental Microbiology* 78(12), 4501–4504.

Freeman, L.A., Carpenter, M.C., Rosenberry, D.O., Rousseau, J.P., Unger, R. and McLean, J.S. (2004). Techniques of water-resources investigations 8-A3. Use of Submersible Pressure Transducers in Water-Resources Investigations. Reston, VA: United States Geological Survey.

Lewis, J. and Eads, R. (2001). Turbidity threshold sampling for suspended sediment load estimation. In Proceedings of the Seventh Federal Interagency Sedimentation Conference. Technical Committee of the Subcommittee on Sedimentation. 25–29 March. Reno, NV. Reston, VA: United States Geographical Survey.

Murdock, A. and McKnight, S.D. (1994). *Handbook of Techniques for Aquatic Sediment Sampling*. Chelsea, Michigan: Lewis Publishers.

Nielsen, D.L., Brock, M.A., Rees, G.N. and Baldwin, D.S. (2003). Effects of increasing salinity of freshwater ecosystems in Australia. *Australian Journal of Botany* 51, 655–665.

Phillips, J.M., Russell, M.A. and Walling, D.E. (2000). Time-integrated sampling of fluvial suspended sediment: a simple methodology for small catchments. *Hydrological Processes* 14, 2589–2602.

Radtke D.B. (2005). Bottom-material samples In: Handbooks for Water-Resources Investigations. Techniques of Water-Resources Investigation. Reston, VA: United States Geological Survey.

Rusydi, A.F. (2017). *Correlation between conductivity and total dissolved solid in various types of water: A review*. Global Colloquium on GeoSciences and Engineering. IOP Conference Series: Earth and Environmental Science 118, 012019.

Schimmelmann, J.P., Nguyễn-Văn, H., Nguyễn-Thuỳ, D. and Schimmelmann, A. (2018). Low cost, lightweight gravity coring and improved epoxy impregnation applied to laminated Maar sediment in Vietnam. *Geosciences* 8(5), 176.

Sherriff, S.C., Rowan, J.S., Melland, A.R., Jordan, P., Fenton, O. and O'hUallacháin, D. (2015). Investigating suspended sediment dynamics in contrasting agricultural catchments using ex-situ turbidity-based suspended sediment monitoring. *Hydrology and Earth System Sciences* 19, 3349–3363.

Shore, M., Jordan, P., Mellander, P-E., Kelly-Quinn, M., Daly, K., Sims, J.T., Wall, D.P. and Melland, A.R. (2015). Characterisation of agricultural drainage ditch sediments along the phosphorus transfer continuum in two contrasting agricultural catchments. *Journal of Soils and Sediments* 16, 1643–1654. doi:https://doi.org/10.1007/s11368-015-1330-0

The European Parliament and the Council of the European Union. (2000). Directive 2000/60/EC of the European Parliament and of the Council of 23 October 2000 establishing a framework for Community action in the field of water policy. *Official Journal of the European Communities* 327, 1–72.

Trost, J.J., Christy, T.M. and Bekins, B.A. (2018). A direct-push freezing core barrel for sampling unconsolidated subsurface sediments and adjacent pore fluids. *Vadose Zone Journal* 17, 180037.

United States Environmental Protection Agency. (2013). *Groundwater Sampling*. Washington, D.C.: United States Environmental Protection Agency Science and Ecosystem Support Division.

United States Geological Survey. (2006). *Techniques of Water-Resources Investigations of the U.S. Geological Survey. Book 9*. Reston, VA: United States Geological Survey.

Webb, B.W., Hannah, D.M., Moore, R.D., Brown, L.E. and Nobilis, F. (2008). Recent advances in stream and river temperature research. *Hydrological Processes* 22, 902–918.

Winter, T.C., Harvey, J.W., Franke, O.L. and Alley, W.M. (1998). Ground water and surface water; A single resource. United States Geological Survey Circular 1139. Reston, VA: United States Geological Survey.

8

Plants

Finding a Slope

Slope can be measured using ranging poles and a clinometer. Ranging poles are simply lightweight posts, which often have bands of color (usually white and red or orange). Poles are typically 2–3 m tall and may have a flag at top. A clinometer is a lens device which incorporates a level calibrated to a scale. This is used to measure the angle of inclination.

1) Partner 1 should take up position at the base of the slope. Use a tripod or simply hold the clinometer at a known height against the ranging pole. Drive a flag or stake at this starting point.
2) Partner 2 should walk 15 m up the slope and hold their ranging pole vertically with the foot on the ground.
3) Partner 1 should look through the clinometer lens with one eye, keeping their other eye open. Align the crosshairs so that the level is at "0."
4) Once you have established the level, tilt the clinometer until the crosshairs align with a marked point (such as the change in band colors) on Partner 2's ranging pole. Record the angle in either degrees or percentage from the scale on the clinometer.
5) Repeat this process every 15 m. You can then plot the slope. Using smaller distance increments will increase the resolution.

Fieldwork Ready: An Introductory Guide to Field Research for Agriculture, Environment, and Soil Scientists,
First Edition. Sara E. Vero.
© 2021 American Society of Agronomy, Inc., Crop Science Society of America, Inc., and Soil Science Society of America, Inc. Published 2021 by John Wiley & Sons, Inc.
doi:10.2134/fieldwork.c8

Randomizing Plots

When running a plot experiment, treatments are randomized to reduce the impact of small-scale differences (such as soil compaction, slope, moisture content, etc.) on the results. Let us imagine that all your replicates of Treatment A were located side-by-side in a certain area of the field. If this area is wetter than another part of the field where the Treatment B replicates are located, your results may be influenced by the moisture content, rather than by the differences between the two treatments. Randomization reduces this effect and can be done easily using Microsoft Excel (Fig. 8.1).

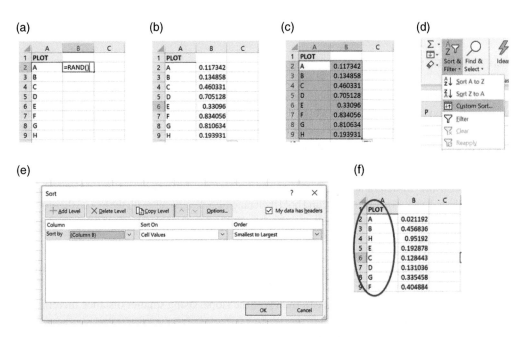

Fig. 8.1 Randomization using Microsoft Excel. (a) List your plots/blocks in column A. Insert the randomize function in column B. (b) Copy down the function. This will generate a random number in each cell. (c) Select both columns. (d) Select "Custom sort" from the sort and filter tab. (e) Sort by column B. (f) This will reorder the plots according to the random number associated with them. *Source:* Sara Vero.

Marking Field Plots

Once you have decided upon your treatments, number of replicates, and your site, you next will need to set out your plots (Fig. 8.2). Your site should be reasonably uniform, although it is acknowledged that some differences in soil texture, slope, past management, porosity, and density, etc. are likely. Sampling these properties in each plot or block can provide contextual information and may be helpful in your interpretation of results. If possible, find out about the previous cropping history and management. Also, ensure that the site is available for the full length of your intended study. If you need data over three growing seasons, then it will be a big problem if the landowner decides to change how that field is managed during the final year! If a single site cannot be secured for the full study, due to crop rotation for example, then comparable sites might be a viable alternative. Each plot must be identical in size and shape. Typically, rectangular plots are used. In order to mark out right-angled plots, you will need to use the Pythagorean Theorem and what is known as the 3–4–5 method (Fig. 8.3):

Fig. 8.2 Grass plots prior to harvesting. *Source:* Sara Vero.

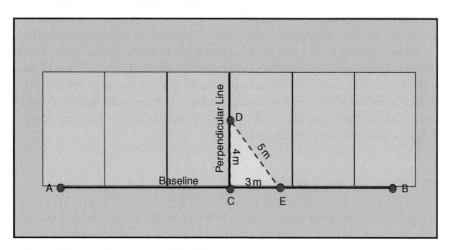

Fig. 8.3 Layout of plots using the 5-4-3 method. *Source:* Sara Vero.

$$a^2 + b^2 = c^2$$

You will need three people to set out your plots, in addition to stakes, a compass, ranging poles (lightweight poles with alternating bands of color to aid visibility), a measuring tape and string or chain.

1) Start by driving a stake where you intend the first corner of your block to be located (Position A).
2) Measure out the intended long edge of your block and drive a second stake (Position B). The line between A and B is your baseline. You can use primary directions (North-South or East-West) to orient this baseline.
3) At the midpoint along the baseline, one person should hold a ranging pole vertically (Position C).

4) Use your compass to identify the perpendicular direction from the baseline midpoint. In other words, if your baseline runs East-West you need to find North. Use your tape to walk 4 m directly in this direction. The second person should hold their ranging pole at this position (D).

5) The third person shall measure out a distance of 3 m along the baseline. This is position E. If your right angle is correct, there should be a distance of 5 m between position D and position E.

6) You now have an initial baseline and right angle. Simply measure out your remaining plots using your tape and mark the corners using stakes or flags.

It is a good idea to tag the GPS coordinates at the corner positions as these will help you locate the plots if markings get damaged or changed. This can easily occur, and preparation in advance can make identifying your plots much easier the following season. You can mark out the lines of the plots using string, fencing, athletic field paint (used on sports pitches), by applying herbicide, or by using hydrated lime. This is powdered calcium hydroxide ($CaOH_2$) and is a strong alkaline, which can cause chemical burns and skin irritation. Wear appropriate personal protective equipment (PPE) including gloves, dust mask, and goggles. This treatment should be applied in thin lines using a narrow spout (watering cans with the rose removed are ideal) along the plotlines. This will kill vegetation and leave an outline of your plots. Keep in mind that over several growing seasons weeds may encroach on the bare soil so annual treatment might be necessary. Weathering and the repeated harvesting of field plots can also cause plot outlines to become worn or faded. Maintenance is therefore vital if you intend to use the plots for a number of years. It is a good idea to schedule this in advance as part of your annual fieldwork plan. Use flags or other markers to indicate the plot numbers and treatment details.

When laying out plots it is important to leave sufficient space for vehicle manoeuvring. Do not encroach upon neighboring plots when applying treatment to or harvesting a plot. Although it is often desirable to keep your field site relatively compact (particularly if on a farmers' land rather than a research station), leaving an alley between blocks or plots will limit impacts on adjacent plots. Consider the axle and tyre width of tractors, mowers, or other machinery that you intend to use on the plots when calculating the width of the alley required. Leave sufficient space at either end of the blocks/plots to allow vehicles to turn.

Applying Treatment to Field Plots

Treatments applied to field plots might include different fertilizer formulations, pesticides, soil amendments, manures, etc. The treatments may vary in their composition or rates. Regardless of these variables, it is vital that treatments are applied accurately (i.e., in the correct amounts), to the correct plots and in an even fashion. In relation to the last point, you do not want to apply 80% of your fertilizer to 30% of the plot! Here are some useful pointers:

- Weigh all treatments prior to going to the field if possible. This should be achievable for most chemical treatments. Bag and clearly label each treatment and which plot it is intended for. If your treatment is a uniform material (e.g., fertilizer granules of a consistent diameter), you could use a container of a known volume to deliver a single dose of treatment. For example, if 100 g of the treatment fills a container of 250 mL, this can be used to dispense your treatment in the field. You should always test this approach in the laboratory by repeatedly weighing test samples.

- Always bring some spare preweighed treatment. This can come in handy if you lose a bag or if one spills during transport.

- If you are applying animal slurry, you may need to weigh the treatment on site (Fig. 8.4). This can occur if slurry is being supplied by the landowner/farmer, or if you are transporting slurry in a

Fig. 8.4 Preparing to apply cattle slurry to grass plots using watering cans. *Source:* Sara Vero.

bulk container such as a tanker and then decanting on-site. Bring a suitably large, calibrated field scale. Ensure that you have a stable platform to place beneath the scale as it will not be stable or accurate if placed on a grass surface. You can use a slurry hydrometer in the field to generally indicate dry matter (DM) and hence to calculate your application rate. The slurry hydrometer is a graduated glass device which is inserted into a container (usually a 1 L jug) of slurry. It will settle at a percentage which indicates the dry matter. This can be related to the nutrient value of the slurry using look-up-tables in agronomic advisory books (e.g., Wall and Plunkett, 2016) or published papers (Zhu et al., 2003). Depending on the DM of your slurry, you can adjust the volume which you apply in order to match your target nutrient content. Alternatively, if you are preparing your slurry treatment in advance, you can directly measure the dry matter in the laboratory before transporting to the field.

- Check twice before you apply a treatment! This seems obvious but it is very easy to mix up your plot numbers or your treatment formulations. Double check your maps and labels before you apply.
- Apply treatments consistently across plots, covering the entire plot area (Fig. 8.5). This takes practice. If applying a granular formulation manually it is best to make several passes over the plot, sprinkling ¼ to 1/3 of the treatment with each pass (Fig. 8.6). It is best to sprinkle a little at a time rather than applying large portions. Take your time. Avoid a very windy day as this can blow treatment onto neighboring plots.
- If applying a liquid formulation using a knapsack/backpack sprayer, ensure that you are using the correct nozzle, that the sprayer is calibrated, and that you are making the appropriate number of sprays per pace (Fig. 8.7). The design of the nozzle will determine the width of the spray and the size of the droplets. Smaller droplets will allow more even coverage but keep in mind that the viscosity of your fluid may influence the suitability of different nozzle apertures. The applied pressure will also influence droplet size. Lower pressure creates larger droplets, while higher pressure creates smaller droplets. Most hand-operated sprayers operate at pressures of 300–600 kPa. Many sprayers include a pressure gauge to help you monitor this and adjust as necessary.

Calibration is used to ensure a consistent volume of treatment is delivered with each spray and across the treatment area. You must first determine the volume delivered with each spray. To do this, first fill the knapsack/backpack with water (there is no need to waste chemical during calibration).

Fig. 8.5 Applying cattle slurry to grass plots. Multiple passes are made to ensure an even distribution.
Source: Sara Vero.

Fig. 8.6 Granular fertilizer can be applied to small plots by hand. Use small handfuls and distribute evenly across the plot.
Source: Sara Vero.

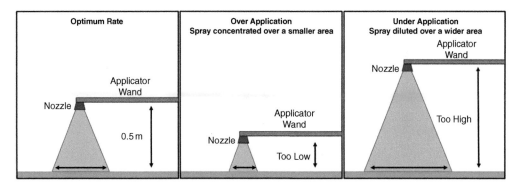

Fig. 8.7 Application of liquid treatment using an applicator wand. The height above ground surface will influence the concentration of application; it is important to be consistent. *Source:* Sara Vero.

Direct the nozzle into a flask or bucket and measure the volume of water released in a single spray. Repeat this several times to find the average.

Next you need to determine the volume of application per area. First measure spray width. This is easiest to do on a dry concrete pavement as the water will show up clearly. Keeping the nozzle at the recommended height above the ground (around 0.5 m in most cases but check the manufacturer's instructions), release a spray of water. Measure the width of the spray. You can then adjust to your desired width using the nozzle. Remember, depending on the width of your plots, you may need to make several passes. Each pass should receive a single spray of treatment, with no overlap. It is essential to hold the application wand at the same height. This keeps the angle of spray and hence, the coverage, the same throughout the treatment.

Once your spray width is correct, mark an area matching the dimensions of your plots. Spray your test area using the intended pressure and ground speed until you have applied the target volume. You may want to practice this several times until you can reliably apply an equal treatment across the plot. A metronome or timer can help you achieve this; you can easily download apps for these on most smartphones.

Phenology

Identification of the growth stage of a crop is known as phenology. This is an essential skill in crop research as it is facilitates application of treatments at specific points and the characterization of response throughout the crop's life cycle. There are various approaches to classification for individual crops, and furthermore, many crops may be assessed using a number of different scales. For example, cereal crops may be characterized using the Zadoks scale (Zadoks et al., 1974 – common in Europe and Australia) or the Feekes scale (Feekes, 1941 – common in the United States). Of course, growth stages differ substantially between species, so you should consider the standard characterization for your species of interest. However, we will briefly discuss how to identify the growth stage of four common species which you might encounter: wheat (Fig. 8.8), corn (Fig. 8.9), and perennial forage grasses (Fig. 8.10), and soybean (Fig. 8.11).

Of course, a plot or row will have many plants so you should select a number of individuals. Their development is used to indicate that of the population as a whole. These individuals should be selected from middle rows or from the center of the plot to minimize edge effects and tagged so that they can be repeatedly measured (e.g., Mahama et al., 2015). Unbiased sampling can be challenging, as selection or rejection of specific plants within the population can occur either based on a cursory visual examination or as the result of an unconscious bias. This can be overcome by sampling plants at a predetermined distance along or within the row, or by sampling every nth plant.

Wheat

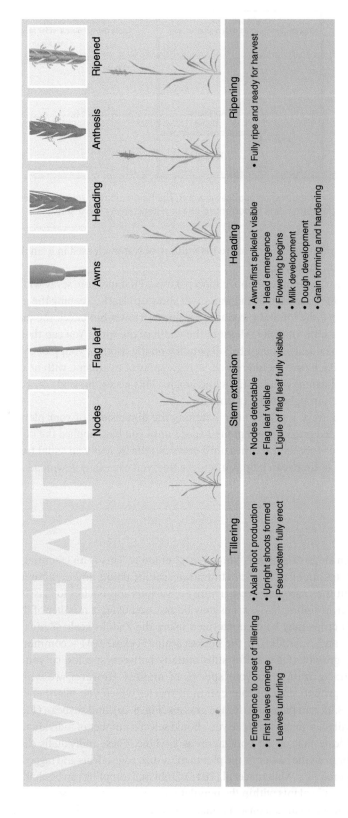

Fig. 8.8 Growth stages of wheat. *Source:* Karen Brey.

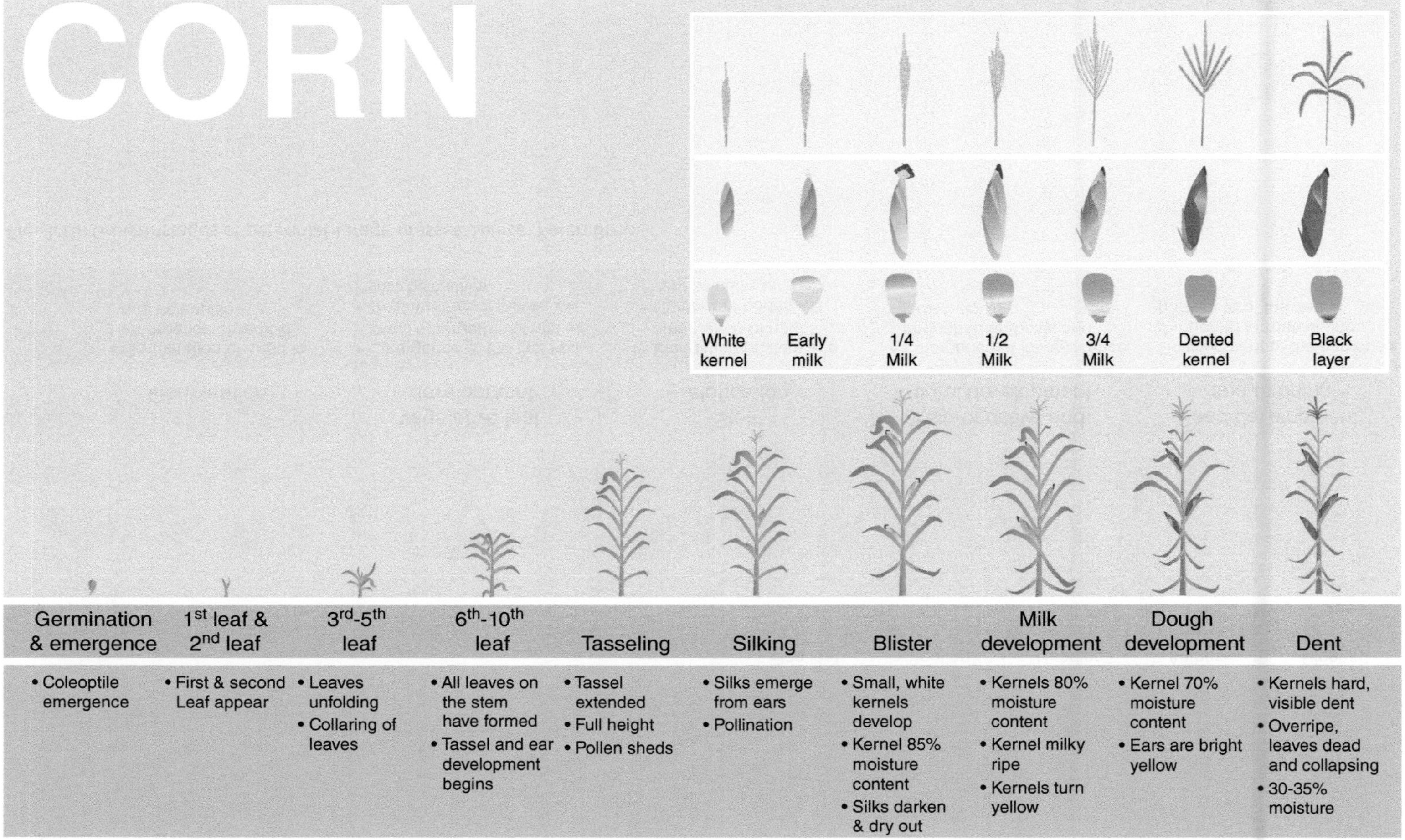

Corn

CORN

Germination & emergence	1st leaf & 2nd leaf	3rd–5th leaf	6th–10th leaf	Tasseling	Silking	Blister	Milk development	Dough development	Dent
• Coleoptile emergence	• First & second Leaf appear	• Leaves unfolding • Collaring of leaves	• All leaves on the stem have formed • Tassel and ear development begins	• Tassel extended • Full height • Pollen sheds	• Silks emerge from ears • Pollination	• Small, white kernels develop • Kernel 85% moisture content • Silks darken & dry out	• Kernels 80% moisture content • Kernel milky ripe • Kernels turn yellow	• Kernel 70% moisture content • Ears are bright yellow	• Kernels hard, visible dent • Overripe, leaves dead and collapsing • 30-35% moisture

White kernel | Early milk | 1/4 Milk | 1/2 Milk | 3/4 Milk | Dented kernel | Black layer

Fig. 8.9 Growth stages of corn. *Source:* Karen Brey.

Perennial Forage Grasses

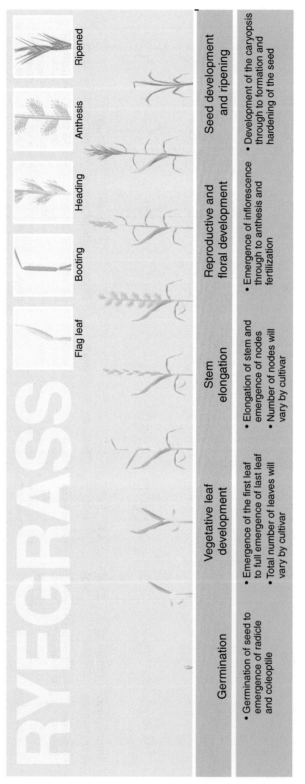

Fig. 8.10 Growth stages of perennial forage grasses. *Source:* Karen Brey.

Soybeans

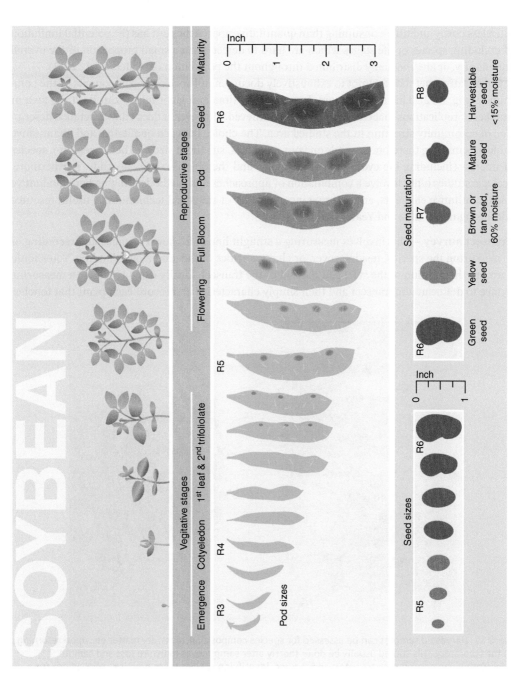

Fig. 8.11 Growth stages of soybeans. *Source*: Karen Brey.

Vegetation Sampling

Vegetative sampling can be broadly described as "qualitative" or "quantitative." Qualitative methods involve recording the presence or absence of a feature, species, or object. A qualitative survey typically is used to determine the composition or distribution of plants over a large area. In this approach, sampling units are spread over a wide (subjective) area and so the intensity is lower than in quantitative approaches. The utility of a qualitative approach is that it allows large areas to be sampled (hopefully representatively), giving a picture of the overall ecosystem. This approach is often less costly and time-consuming than quantitative approaches but has the potential limitation of excluding species or life stages which are hard to detect, form a small proportion of the overall community, or are unequally distributed throughout the population (Golodets et al., 2013).

Quantitative methods attempt to exhaustively document all species or components of the community (Fig. 8.12). This may be more demanding of time and labor which can limit the area or number of replications that can be realistically achieved. However, it does allow a detailed description of community structure in the studied area. The choice between qualitative and quantitative ecologic sampling therefore depends on your resources, site (both size and heterogeneity), species of interest (including life cycle and distribution), and the objective of your study. Furthermore, ecologic studies often involve a combination of approaches and methods bringing both qualitative and quantitative elements, and a strong incorporation of statistical techniques. A useful resource addressing this is Glaz and Yeater (2017).

Transect survey – This involves measuring a straight line of a known distance and recording or measuring the species, residue cover, or characteristics of the crop along that line. You should record the position of the start and the end of the transect, ideally using GPS. Use a measuring tape to delineate the transect and then simply characterize and record each plant that touches

Fig. 8.12 Harvested samples can be assessed for species composition, mass, dry matter etc. upon returning to the laboratory. This should usually be done shortly after sampling, as moisture loss and sample degradation can occur during transport and storage. Identifying species takes practice - seek advice from experienced researchers and use an identification key. *Source:* David Jaramillo.

that line at any point (line survey) or at prescribed intervals (e.g., every meter) along the line (point survey). The shorter the interval or the greater the number of sample points, the greater the resolution of the survey.

Transect surveys can be used to indicate changes in composition over time. In such an approach, the line is permanently marked either by recording the start and finish locations or by installing flags or stakes so that the transect points can be returned to. This allows the same line to be sampled across time (e.g., Capers, 2000). Transect approaches can also be combined with plot or quadrat methods. In these cases, plots or quadrats are implemented at points along the transect (e.g., Ratajczak et al., 2014).

Quadrat – Quadrats can be used to sample or survey vegetation, particularly where cover, density, or mass is of interest. In this approach, a rigid frame of known area (the quadrat) is placed either randomly or at a selected location in the landscape, along a transect or within a plot. Quadrats of 0.5×0.5 m and 0.25×0.25 m are common, although other sizes and shapes are available. Quadrats can be quite simply made using timber. Each quadrat may be further subdivided using string or wire into smaller grids to allow more accurate estimation of percentage cover (Fig. 8.13).

Percentage cover – This is used to determine the area covered by all plants present (and inversely, bare soil) or by a particular species. This is done using a Daubenmire frame which is similar to a quadrat measuring 0.2 m $\times$ 0.5 m (Daubenmire, 1959; Coullouden et al., 1999, Bonham et al., 2004). Place the frame on the ground and view from directly above. The area within the frame covered by canopy vegetation is estimated by eye and assigned to one of six classes (Table 8.1). Presence of species with different heights can

Table 8.1 Daubenmire cover classes.

Class	Cover (%)	Midpoint (%)
1	0–5	2.5
2	6–25	15.5
3	26–50	37.5
4	51–75	62.5
5	76–95	85
6	96–100	97.5

Fig. 8.13 This researcher is using a quadrat which has been subdivided into smaller squares to characterize plant composition at an upland research site. *Source:* Noemi Naszarkowski.

lead to overlapping coverage, which in dense swards may result in a coverage exceeding 100% when all species within the frame are summed. You may also take photographs from above the frame for analysis when you return from the field. Record the class of each quadrat (replicate) and then calculate the average cover by adding the mid-point of each replicate and dividing by the total number of quadrats measured.

Let us try an example (Table 8.2). You are studying the effects of low-intensity sheep grazing on the species composition of unmanaged hillslopes. You set up a 50 m transect, and measure covers every 10 m.

The hillslope is therefore assigned a Daubenmire cover class of five. The use of cover classes helps to overcome variation in estimates which may occur due to the visual nature of assessment. Such variations may be particularly pronounced when more than one person is conducting measurements. Sorting into classes thereby encourages a more consistent scoring.

Table 8.2 Example of Daubenmire cover class calculation.

Replicate	A	B	C	D	E	F
Distance from start (m)	0	10	20	30	40	50
Cover (%)	79	83	62	85	70	90
Midpoint (%)	85	85	62.5	85	62.5	85
$85 + 85 + 62.5 + 85 + 62.5 + 85 = 465$			$465 \div 6$ (number of plots) $= 77.5$			

Canopy analysis instruments – Advances in gap fraction analysis and radiation sensors has allowed the development of handheld devices that indirectly calculate foliage density and canopy characteristics (Fig. 8.14). An increasing range of these instruments are available commercially, and

Fig. 8.14 Handheld canopy analysis devices can non-destructively measure plant density. *Source:* Jesse Nippert.

there is a burgeoning literature comparing their performance relative to both other instruments and traditional methods (Welles and Cohen, 1996; Wilhelm et al., 2000; Cescatti, 2007) These devices allow nondestructive analysis and deliver rapid measurements in the field. Wilhelm et al. (2000) examined several commercially available devices and found that each of the tested devices delivered adequate measurements when properly used according to the manufacturer's instructions. Those authors advised that the user carefully considers the objectives of their study before purchasing equipment and closely study the user manual. This is sound advice for any fieldwork!

Another relatively recent tool is smartphone apps which estimate leaf area index based on color ratios (e.g., Patrignani and Ochsner, 2015). Apps incorporate the empirical, mathematical approach of desktop image analysis software with ease of measurement in the field. Furthermore, they may be considerably less expensive. As the quality of smartphone cameras increases, the use of apps is likely to become more widespread, and their performance will further improve. Care must be taken to avoid user error. While the specific instructions differ depending on app and smartphone, in general, the measurement should be taken at a consistent, specified height above the crop and the camera lens should be level.

A benchtop leaf area meter can be used to evaluate individual leaf dimensions back at the laboratory if samples are stored in moist, cool conditions. These devices flatten the leaves and so may enable an improved measurement of total leaf area for curled or crinkled leaves than would otherwise be achieved.

Composition and density – Composition is the proportion of species observed within your quadrat based on the cover, density or weight each species represents (Fig. 8.15). For each species, you should divide the percentage cover (for example) by that of the total cover in the frame. The total composition should always add up to 100%. Let us imagine a mixed species grassland on a dairy farm (Table 8.3). The total cover is 72% and includes perennial ryegrass, timothy, and red clover.

Table 8.3 Example of cover and composition within a quadrat

Species	Cover (%)	Composition
Perennial ryegrass	53	0.74
Timothy	10	0.14
Red clover	9	0.12

Fig. 8.15 These researchers are taking measurements and identifying plant composition. *Source:* Jesse Nippert.

Similarly, density refers to the absolute number of a certain species within the sampling area (absolute density) or the percentage of the community represented by a particular species (relative density).

Frequency – This is the number of times a species is encountered within the number of quadrats sampled. Naturally, this varies so having a sufficiently high number of quadrats is essential to gain a statistically sound representation. Measurement of frequency can be affected by the size of your quadrat, as larger quadrats may be more likely to include a given species.

Destructive sampling – In this approach, the vegetation within the quadrat is sheared and removed (Fig. 8.16). It can then be analyzed for mass, dry matter, nutrient content, species, etc. back at the laboratory. Motorized or hand-shears can be used for this. If using motorized shears, wear protective gloves as they can easily and severely cut fingers.

Fig. 8.16 These grass plots have been harvested and surplus vegetation removed. *Source:* Sara Vero.

Grass yield – Destructive sampling consists of harvesting a known area (a field, plot, or quadrat) and measuring its dry matter (see vegetation sampling). These samples may also be analyzed for nutritional content. However, you may wish to calculate the yield of a growing grass crop without destructively harvesting an entire plot or field. By not disturbing or removing the crop, you can monitor growth over time. Crop height alone may be useful, but that approach has a major limitation in that it does not indicate dry matter. As a result, it will not be possible to calculate the capacity of that crop to support livestock grazing or how suitable it is for silage or hay production.

The rising plate meter is a simple, effective, and relatively rapid method of assessment and has gained popularity both in research and in farming. The rising plate meter is essentially a staff (about the length of a walking stick or crutch), which has a lightweight metal or plastic disk/plate which can

Fig. 8.17 Rising plate meters can be used to provide a non-destructive measurement of forage yield. *Source:* David Jaramillo.

move freely when the foot of the staff is pressed against the soil (Fig. 8.17). The height to which the plate rises will be influenced by both the height and the density of the crop. You should walk the field in a W-shape, just as you would for soil sampling, and press the staff to the ground every few paces. It is vital that you bring the meter straight up and down to make an accurate measurement. The meter will take a rolling tally of crop height and also record the number of measurements you take. The rising plate meter is designed not just for research but also as a farm management tool used to evaluate forage yield field or paddock scale. Recommendations differ depending on the manufacturer, but in general, you should take approximately 30 measurements per paddock or 1 measurement every 25 paces. Of course, paddock sizes vary so it may be sensible to take a similar approach to soil sampling for nutrient analysis using parcels of <2 ha 5 acre^{-1}. As with soil sampling, you may sample areas of the field which have significantly different features separately. For example, if one area of the field is particularly damp, it is likely to have different growth patterns than the remainder. It would be sensible to measure this area separately rather than skewing the reading for the entire field.

At the end of your measurements, you will divide the cumulative plate meter reading and then divide by the number of readings. This is converted to yield by multiplying by a calibration equation provided by the manufacturer (equations vary by manufacturer, vegetation type (i.e., species and phenology), production system norms and national standards). A more accurate estimate for an individual paddock can be created by measuring the dry matter (see below) and incorporating into the calibration equation.

Dry matter – Dry matter (DM) provides a more reliable indicator of the nutrient concentration and energy value of vegetation than wet weight. While the temperatures used can vary depending on the type of plant being examined, the most common method of assessment is to weigh a sample and dry in an oven at 65 °C (Reuter and Robinson, 1997) until weight remains constant. Subsequent to drying, the sample is reweighed, and DM is calculated as a percentage of wet weight. Drying at above 80 °C can result in losses due to nitrogen volatilization, while drying at ambient temperatures can incur losses via respiration. Microwaves can also be used, although care should be taken that the sample does not burn. If samples are allowed to burn, constituents other than water are removed so the final value will not be accurate. Take care to tare the scale carefully prior to drying.

Two excellent and thorough handbooks on ecological sampling are Elzinga et al. (1999) and Krebs (2008).

Root Sampling

Although treatments may directly affect the morphology and function of plant roots, the above-ground portion of the plant is frequently given greater attention in research. This may not only reflect in part the importance of shoots, leaves, and fruits from a commercial and agronomic perspective but also the relative ease with which these components can be sampled. This should not discourage you, however. Recent developments in non-invasive techniques such as X-ray tomography or root capacitance offer huge potential to increase the accuracy and ease with which the phenotype and functioning of roots can be observed. Even without new technologies, there are some techniques which can be used to directly sample roots from plot or pot experiments.

Excavation followed by careful washing to remove soil can allow entire root systems to be maintained intact. This enables detailed analysis of morphology but does require care and time so as not to damage the delicate structure. For plants with relatively small root systems, coring is an option. Similar to soil coring for density or structural analysis, this involves driving a ring of known volume into the ground and then extracting (Smit et al., 2000). The dimensions may vary, but cores of <10 cm diameter are common (Eshel and Beeckman, 2013). Successive cores are extracted at various depths. Alternatively, a soil block ("monolith") can be excavated by digging or cutting around the entire root span. The block is then lifted upward and boarded on each side using card or light plywood and tape. This will help keep the block intact during transport to the laboratory. Be aware that soil type will influence the ease of operation. Sandy soils or those with low cohesion may fall apart as they are removed, and roots can be damaged during this. Conversely, soils with a high clay content or a large proportion of stones may be difficult to cut through and heavy to move (Eshel and Beeckman, 2013).

After sampling, the soil surrounding the roots must be gently removed so that analysis can begin. This can be a rather tedious and time-consuming task. Roots will begin to decay after sampling, so if you cannot wash within 2–5 d (if storing at 15–25 °C), then freezing is an appropriate storage method. Samples must be thawed for a day prior to washing.

Coarse soils may be easily removed by hand or by gently sprinkling with water. As clay or organic matter increases, washing may become more difficult. Samples can be soaked in water and gently agitated by hand to encourage roots to float upward, away from the denser soil particles. Soaking in a mild dispersant such as sodium hexametaphosphate ($100\,g\,L^{-1}$) can also be used (Thivierge et al., 2015).

Cleaned roots can then be analyzed using a wide range of techniques, not limited to length, diameter, mass (and ratios of those properties), chemical composition, etc. Root scanning has become increasingly popular. This approach uses high-resolution scanners and image processing software such as WinRHIZ, RootScan (Burton et al., 2012), or others. These methods allow more detailed characterization of root architecture.

Plant Height

The recent development and increased availability of unmanned aerial vehicles (UAVs), commonly referred to as "drones," has made high spatiotemporal and nondestructive measurement of crop (and other vegetation) development possible over greater scales and with lower costs than manual approaches. This seems likely to become progressively widespread as the accuracy of sensors and speed of data processing become greater. UAV data already has become an essential tool in precision agriculture (Anthony et al., 2014; Maes and Steppe, 2018). However, at pot, plot, and block-scale, traditional methods are still crucial. Plant height is a basic measurement you are likely to use. For many species of interest, height will vary across the crop. Furthermore, in a mixed sward or community, there will be a number of species or varieties. The rising plate meter discussed earlier offers one approach, in which a mean measurement is achieved incorporating an unspecified number of individual plants. This is similar to composite sampling in some ways and gives a good estimate of overall forage grass development. In many cases, and for other species, more detailed measurements are necessary. As discussed in the Phenology section, several individuals can be tagged and measured repeatedly. Maximum height may not be the optimum measurement, depending on the objective or species. Measurements may also be taken to specific features such as nodes, internodes, ears, or the base of the tassel, which may be more informative than maximum height. It is typical to exclude the inflorescence from height measurements. Always record the method which you use accurately.

A measuring rod or meter stick can be used for plants that are not too tall to accurately identify the increment. The measuring rod should be held upright with the base on the soil surface. You may need one person to hold the rod and the other to read the measurement. Individual small plants or those in early growth stages may be pulled up and measured by laying flat alongside a ruler or tape measure. If this approach is used, you should measure from the location that the shoot emerges above the soil surface.

Tree height – This can be challenging for trees which are several meters tall, so trigonometry may replace the direct measurements used for smaller plants (see later section) (Fig. 8.18). You will need a clinometer. Take a position with a clear line of sight at a known distance from the tree. This is d (m). Use the clinometer to identify the angle from your position to the tree-top (α) and from your position to the tree base (β). Height (H) is calculated as:

$$H = d \times \left[\mathrm{Tan}(\alpha) \times \mathrm{Tan}(\beta) \right]$$

This approach works best when the angle is between 30° and 45°, you may need to adjust your distance (d) from the tree to accommodate this.

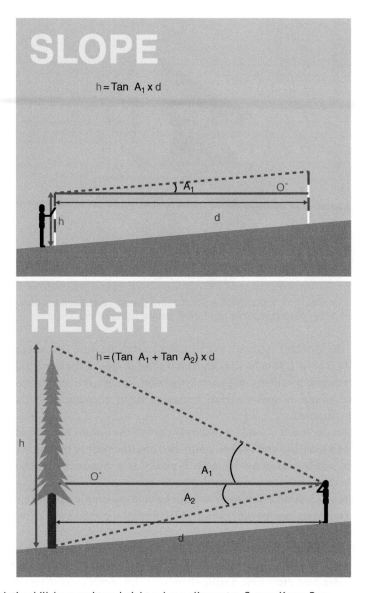

Fig. 8.18 Calculating hillslope and tree height using a clinometer. *Source:* Karen Brey.

Fig. 8.19 Tree diameter should be measured at 4.5 feet above ground level. Keep the tape or dendrometer level and ensure that it is in contact with the trunk at all points. *Source:* Krista Keels.

Tree diameter – The standard approach to tree diameter is to measure Diameter at Breast Height (DBH) (Fig. 8.19). This is 1.37 m or 4.5 ft high on the uphill side of the tree trunk. Such a standardized measurement is useful in comparing across sites, species, research teams, etc. Of course, trees often have unique or asymmetrical trunks so slight adaptations to the measurement approach may be required.

- **Calipers** – This approach is used for relatively narrow trees or saplings. Holding the calipers horizontally and level, the arms will be positioned on either side of the trunk and the distance between them will be shown by the scale. Typically, two measurements should be taken at right angles to one another and averaged.
- **Tape** – This is a very simple approach and can be done by simply passing a tape around and ensuring that it is snug against the trunk at DBH. Record the length on the tape and divide by π (3.14) to indicate diameter.
- **Band dendrometer** – A dendrometer is a slim metal or plastic band which is passed flush around the trunk at DBH and fastened using a spring. As the tree grows, the spring will stretch. Some dendrometers have a scale showing 1 mm increments so you can measure the length that the spring is stretched over time. Alternatively, you can measure spring length using calipers. Make certain to record the initial spring length and the date of installation. As tree growth is typically slow, dendrometers can remain in situ for many years. You should inspect regularly for damage such as corrosion or wear. If the initial spring becomes too small for the tree, you can replace it with a longer one or simply link another spring. Datalogging dendrometers are also available which incorporate tensiometers. These allow high accuracy and temporal resolution and have applications where diurnal shrink-swell is of interest.

Yield

Forage Crops

For some crops such as grassland, measurement of yield can be done by harvesting a known area such as a quadrat or plot, weighing the total sample and then calculating the dry matter by drying a subsample (see "Vegetation Sampling") (Fig. 8.20). For large plots, this may require a ride-on mower or grass harvester (Fig. 8.21). The former option is ideal for smaller plots and can be transported to distant sites on a trailer. However, you must weigh the grass by emptying the hopper into a weigh-boat after each plot, so it is best to have a partner or two to help you. Make sure you collect all grass from the hopper each time and that the scale is on a flat, level surface. Accurate collection and measurement can be difficult in very windy conditions. Wind screens or temporary shelters may be helpful. Grass harvesters specially designed for research plots are also available. These are capable of on-board weighing; however, it may not be feasible to transport to various locations outside of dedicated research farms. When designing your plot experiment, you should consider how harvesting will be conducted. After sampling, the edges of your plot, or the entire outside of your quadrat if you use a subsampling approach, may need to be trimmed to the same post-harvest height. This ensures that the following sampling round won't be biased by the effects of preceding round. If a large area must be trimmed, you can use a grass mower or tractor-mounted topper to cut to a level height. Adjust the mowing blades to the desired height and check using a ruler. A hand-held shears can be used to maintain small plots or pots. In that case, you should measure the vegetation height as you go to ensure that you keep a consistent height. Unless it is part of your experimental design, you should remove this surplus vegetation from the plots and dispose of.

Fig. 8.20 Herbage samples should be bagged, labeled and weighed as soon as possible before moisture loss has occurred. *Source:* Sara Vero.

Fig. 8.21 You should choose mowing or harvesting equipment based on the dimensions of the plot, your crop and ground conditions. Here we see a motorized mower, a crop harvester, and a ride-on grass harvester. *Sources:* (a), (c) Sara Vero and (b) David Jaramillo.

Depending on the size of your plot, there may be a relatively large total amount of material harvested. This can be impractical to dry and process and subsampling for laboratory analysis is common, although you should typically still weigh the entire wet mass of vegetation. The subsample should be representative of the entire plot, that is, have a similar ratio of stem to leaf, small to large pieces, and distribution of species. You can address this by taking several small grab samples from the total vegetation mass. Take samples from the top, middle, and bottom of the heap and then composite them (Fig. 8.22).

Dry matter (DM) of grass and forage samples can be measured by drying in a forced air oven at 65 °C until constant weight is reached, usually within 24–48 h. Dry matter is expressed as a percentage of the original "wet" weight. Be aware that the wet weight of grass can fluctuate throughout the day depending on weather conditions.

Fig. 8.22 These researchers are collecting herbage from the harvester and sub-sampling for analysis. Note that they are also wearing ear protectors while working with noisy machinery. *Source:* David Jaramillo.

1) Weigh empty weigh-boat or tray.
2) Add "wet" grass sample; weigh.
3) Dry until constant weight is reached.
4) Reweigh weigh-boat and sample.

$$\text{Weigh} - \text{boat} = 200\,\text{g}$$

$$\text{Weigh} - \text{boat} + \text{wet sample} = 350\,\text{g, therefore, wet sample} = 150\,\text{g}$$

$$\text{Weigh} - \text{boat} + \text{dry sample} = 250\,\text{g, therefore, dry sample} = 50\,\text{g}$$

$$50 \div 150 = 0.33 \times 100 = \textbf{33\% DM}$$

Once you know the DM of the sample, you can calculate yield per acre. Let us work through an example.

$$\text{DM} = 33\%$$

The wet weight of a sample from a 0.5×0.5 quadrat ($0.25\,\text{m}^2$) is $280\,\text{g}$ (or $0.280\,\text{kg}$)

There are $10,000\,\text{m}^2\,\text{ha}^{-1}$; that equates to $40,000$ quadrats.

$$0.280\,\text{kg} \times 0.33 \times 40,000 = \textbf{3696\,kg DM ha}^{-1}\left(\text{or } 3.7\,\text{t DM ha}^{-1}\right).$$

Grain Crops

The final yield of a grain crop is frequently a key indicator of the overall effect of the treatment. This should be calculated as the grain mass per unit area for example, t ha^{-1}, when harvested at full maturity. This can be converted to volume per ac^{-1} or ha^{-1} by multiplying by a standard bushel weight (which corresponds to the methodology used at point of sale for commercial crops), or by the measured test weight (see "Seed Mass" below).

However, you may wish to assess crop development at various growth stages, in which case measurement of final yield is not yet possible. Alternatively, after harvesting a mature crop, you may wish to disentangle which growth stages were most influenced by the treatment. This may not be wholly clear based on gross yield alone. Investigation of the various components of yield can support a more comprehensive analysis. There are three components of crop yield; ears/pods/heads per plant, seeds per ear, and seed mass. Each component develops during a different stage in the crop life cycle. Measuring each of these can thus help you determine the influence of a treatment at different growth stages.

First let us consider obtaining your sample. For treatments which are applied over entire rows, a harvester can be used to collect a bulk sample. It is common for harvesters to have on-board yield monitors which measure the mass flow of grain entering the hopper. These systems can be calibrated using direct measurement of grain mass. More detailed yield analysis on harvested bulk samples can be conducted by subsampling from the hopper or grain bin. Alternatively, grab samples from rows or plots can be taken manually or using hand-held harvesters. Remember to remove any guard rows or discard area prior to sampling (see Chapter 2).

Ears, pods, or heads per plant or per area – This is a key indicator as the number of ears (or pods or heads) will directly constrain the potential yield of the plant. Different species and varieties within those species are capable of producing greater or lesser numbers of ears/pods/heads, and this can be influenced by water and nutrient availability throughout crop establishment. This can be assessed by selecting a representative number of stalks and calculating the average number per plant. To avoid bias (such as preferential selection of particular plants either consciously or inadvertently), sampled plants can be randomly selected (e.g., every n^{th} stalk). When determining yield on a field or row scale, ears per area becomes a more useful measurement. You can simply count the number of ears within a quadrat. To convert between measurements per plant and area, you can multiply ears per plant by the number of plants in a given area. It is important to remember that not all tillers develop ears and that not all ears fully develop. This measurement is therefore most closely related to yield when it is taken near to harvest.

Kernels or seeds per ear – This is indicated by counting the kernels or seeds in a representative number of ears/heads/pods. Depending on your crop you may need to dissect the spikelet to count the individual seeds inside.

Seed mass – For rough estimates, default values of grain or kernel weights are sometimes used. However, in reality, seed mass can vary greatly, even by hundreds of seeds per pound. Measurement of seed mass will therefore deliver far more accurate results. This is relatively straightforward. Mass can vary among seeds or kernels as a result of variable grain fill, maturity of individual plants within the plot and differences in the size of seeds within a single ear or head. To ensure representativeness, either "thousand grain weight" or "test weight" is used and corrected for moisture. The former is the weight of 1000 seeds (in g). Seed counters which incorporate scales and high-speed dispensers are also commonly available.

Test weight is similar; it is the weight of a known volume of seeds. This can be influenced by the packing efficiency of seeds. Seeds which can pack closely together due to the shape of their kernels will therefore have a higher rest weight than seeds which do not pack closely, and more void space occurs between them. While thousand grain weight is influenced by the density of the seeds, test weight reflects both density and shape. In the United States, the standard volume used for measuring test weight is one bushel (2150.42 in.^{3} or 8 gallons). The USDA approves certain equipment for test weight measurement. These items include a scale, hopper, a stand, quart kettle, overflow pan, and a stroker. Detailed instructions for measurement are available in the Grain Inspection Handbook, Book II. Grain Grading Procedures (USDA, 2013). In summary,

- Place the quart kettle on the scale and zero.
- Fill the hopper on a stand with your grain sample.
- Place the kettle beneath the hopper and dispense enough grain to fill the kettle.
- Use the stroker to gently sweep excess grain from the kettle so that it is just full.
- Weigh the full kettle using the scale.

Test scales are not like regular laboratory balances. They report a value which is 32 times the actual weight applied. This is because a bushel contains 8 gallons, so as the kettle has a volume of 1 quart it would take 32 kettles to fill one bushel. This means that filling the kettle precisely will have a significant influence on the accuracy of the measurement.

Now that we understand the three yield components, let us work through an example. You have a plot study on winter wheat. You have 500 ears per m^2, with an average of 40 grains per ear. The thousand grain weight is 45 g.

$$500 \text{ ears m}^2 \times 40 \text{ kernels ear}^{-1} = 20,000 \text{ kernels m}^2$$
$$20 \text{ thousand grain} \times 45 \text{ g thousand grain}^{-1} = 900 \text{ g m}^2$$

There are 10,000 m^2 per hectare so that equals 9000 kg or 9 t ha^{-1}.

References

Anthony, D., Elbaum, S., Lorenz, A. and Detweiler, C. (2014). On crop height estimation with UAVs. In: 2014 IEEE/RSJ International Conference on Intelligent Robots and Systems, 14–18 Sept. 2014, Chicago, IL.

Bonham, C.D., Mergen, D.E. and Montoya, S. (2004). Plant cover estimation: A contiguous Daubenmire frame. *Rangelands* 26(1), 17–22.

Burton, A.L., Williams, M., Lynch, J.P. and Brown, K.M. (2012). RootScan: Software for high-throughput analysis of root anatomical traits. *Plant and Soil* 357(1–2), 189–203.

Capers, R.S. (2000). A comparison of two sampling techniques in the study of submersed macrophyte richness and abundance. *Aquatic Botany* 68, 87–92.

Cescatti, A. (2007). Indirect estimates of canopy gap fraction based on the linear conversion of hemispherical photographs: Methodology and comparison with standard thresholding techniques. *Agricultural and Forest Meteorology* 143(1–2), 1–12.

Coullouden, B., Podborny, P., Eshelman, K., Rasmussen, A., Gianola, J., Robles, B., Habich, N., Shaver, P., Hughes, L., Spehar, J., Johnson, C., Willoughby, J. and Pellant, M. (1999). *Sampling Vegetation Attributes*. Interagency Technical Reference. Washington, DC: United States Department of Agriculture Forest Service & United States Department of the Interior Bureau of Land Management.

Daubenmire, R.F. (1959). A canopy-cover method of vegetational analysis. *Northwest Science* 33, 43–46.

Elzinga, C.L., Salzer, D.W. and Willoughby, J.W. (1999). *Measuring and Monitoring Plant Populations. Bureau of Land Management Technical Reference 1730-1*. Denver, CO: Bureau of Land Management.

Eshel, A. and Beeckman, T., (eds). (2013). *Plant Roots: The Hidden Half*. 4th edition. Boca Raton, FL: CRC Press.

Feekes, W. (1941). De tarwe en haar milieu [Wheat and its environment]. (In Dutch and English). *Verslagen van de Technische Tarwe Commissie* 17, 523–888.

Glaz, B. and Yeater, M., (eds). (2017). *Applied Statistics in Agricultural, Biological, and Environmental Sciences*. Madison, WI: ASA, CSSA, and SSSA.

Golodets, C., Kigel, J., Sapir, Y. and Sternberg, M. (2013). Quantitative vs qualitative vegetation sampling methods: A lesson from a grazing experiment in a Mediterranean grassland. *Applied Vegetation Science* 16, 502–208.

Knott, C.A. and Lee, C. (2016). *Identifying Soybean Growth Stages*. Lexington, KY: University of Kentucky, College of Agriculture, Food and the Environment.

Krebs, C.J. (2008). *Ecology: The Experimental Analysis of Distribution and Abundance*. 6th edition. London: Benjamin Cummings.

Maes, W.H. and Steppe, K. (2018). Perspectives for remote sensing with unmanned aerial vehicles in precision agriculture. *Trends in Plant Science* 24(2), 152–164.

Mahama, G.Y., Prasad, M.V.V., Roozeboom, K.L., Nippert, J.B. and Rice, C.W. (2015). Response of maize to cover crops, fertilizer nitrogen rates and economic return. *Agronomy Journal* 108(1), 17–31.

Moore, K.J., Moser, L.E., Vogel, K.P., Waller, S.S., Johnson, B.E. and Pedersen, J.F. (1991). Describing and quantifying growth stages of perennial forage grasses. *Agronomy Journal* 83, 1073–1077.

Patrignani, A. and Ochsner, T. (2015). Canopeo: A powerful new tool for measuring green canopy cover. *Agronomy Journal* 107(6), 2312–2320.

Pérez-Harguindeguy, N., Díaz, S., Garnier, E., Lavorel, S., Pooter, H., Jaureguiberry, P., Bret-Harte, M.S., Cornwell, W.K., Craine, J.M., Gurvich, D.E., Urcelay, C., Veneklaas, E.J., Reich, P.B., Poorter, L., Wright, I.J., Ray, P., Enrico, L., Pausas, J.G., de Vos, A.C., Buchmann, N., Funes, G., Quétier, F., Hodgson, J.G., Thompson, K., Morgan, H.D., ter Steege, H., van der Heijden, M.G.A., Sack, L., Blonder, B., Poschlod, P., Vaieretti, M.V., Conti, G., Staver, A.C., Aquino, S. and Cornelissen, J.H.C. (2013). New handbook for standardised measurement of plant functional traits worldwide. *Australian Journal of Botany* 61(3), 167–234. http://dx.doi.org/10.1071/BT12225

Ratajczak, Z., Nippert, J.B. and Ocheltree, T.W. (2014). Abrupt transition of mesic grassland to shrubland: Evidence for thresholds, alternative attractors and regime shifts. *Ecology* 95(9), 2633–2645

Reuters, D. and Robinson, J.B., (eds). (1997). *Plant Analysis: An Interpretation Manual*. 2nd edition. Clayton, Australia: CSIRO Publishing.

Ritchie, S.W., Hanwey, J.J. and Benson, G.O. (1993). *How a Corn Plant Develops*. Spec. Rep. 48 (revised). Ames, IA: Iowa State University of Science and Technology: Cooperative Extension Service.

Smit, A.L., Bengough, A.G., Engels, C., van Noordwijk, M., Pellerin, S. and van de Geijn, S.C. (2000). *Root Methods: A Handbook*. Amsterdam: Springer.

Thivierge, M-N., Angers, D., Chantigne, M.H., Seguin, P. and Vanasse, A. (2015). Root traits and carbon input in field-grown sweet pearl millet, sweet sorghum, and grain corn. *Agronomy Journal* 108(1), 459–471.

United States Department of Agriculture, Federal Grain Inspection Service. (2013). *Grain Inspection Handbook: Book II Grain Grading Procedures*. Washington, DC: Federal Grain Inspection Services.

Wall, D.P. and Plunkett, M. (2016). *Major and Micro-Nutrient Advice for Productive Agricultural Crops*. Johnstown Castle, Wexford, UK: Teagasc.

Welles, J.M. and Cohen, S. (1996). Canopy structure measurement by gap fraction analysis using commercial instrumentation. *Journal of Experimental Botany* 47(302), 1335–1342

Wilhelm, W.W., Ruwe, K. and Schlemmer, M.R. (2000). Comparison of three leaf area index meters in a corn canopy. *Crop Science* 40, 1179–1183. doi:https://doi.org/10.2135/cropsci2000.4041179x

Zadoks, J.C., Chang, T.T. and Kanzok, C.F. (1974). A decimal code for the growth stages of cereals. *Weed Research* 14, 415–421. doi:https://doi.org/10.1111/j.1365-3180.1974.tb01084.x

Zhu, J., Zhang, Z. and Ndegwa, P.M. (2003). Using a soil hydrometer to measure the nitrogen and phosphorus contents in pig slurries. *Biosystems Engineering* 85 (1), 121–128.

9

Animal Techniques

Field research that involves animals in any way, either wild or domestic, comes with additional ethical requirements (Russow and Theran, 2003). Research in relation to animals has a somewhat contentious history and may be perceived as harmful or unethical by certain groups or individuals. However, ecological research is crucial to understand populations, their health, and biosecurity. Furthermore, "indicator" species can be used to signal the quality or health of an environment and can alert us to the impact of anthropogenic activities or the success of mitigation measures. We will not discuss animal testing of any sort here, as our focus is field research such as ecological monitoring and survey.

Live-Catch Trapping

Live-catch traps allow animals to be captured, examined, and released (Fig 9.1) and if done correctly, should not cause any lasting or significant harm to the target species. For small to medium

Fig. 9.1 Always consider the safety and welfare of any animals which could be effected by your fieldwork. *Source:* Krista Keels.

Fieldwork Ready: An Introductory Guide to Field Research for Agriculture, Environment, and Soil Scientists, First Edition. Sara E. Vero.
© 2021 American Society of Agronomy, Inc., Crop Science Society of America, Inc., and Soil Science Society of America, Inc. Published 2021 by John Wiley & Sons, Inc.
doi:10.2134/fieldwork.c9

mammals and for some bird species (especially corvids) cage traps can be used. Trapping is an increasingly rare skill and will be new to many researchers. Excellent overviews of trapping for animal research are provided by Powell and Proulx (2003) and Sikes et al. (2016).

Some key principles to consider are the following:

- Always obey the law. Look up your national and state wildlife regulations. Failure to adhere to these regulations will leave you personally vulnerable to prosecution, which may have severe financial implications and even the risk of imprisonment. If permits are required to trap particular species, or during certain times or areas you must obtain these in advance. In the United States, the Animal Welfare Act (United States Code Annotated, 1996) legislates for animals used in research with the exception of field studies that strictly observational and do not involve "an invasive procedure, harms, or materially alters the behavior of an animal under study." Carefully consider your fieldwork in light of this law. Detailed discussion of this legislation is provided by Mulcahy (2017).
- Proposals for research funding that involves animals in any capacity will typically include an ethics statement or questionnaire. You must consider the implications of your work for the animals in question, and the details of your plan as they pertain to the welfare of those animals before commencing any research project. Remember, there may be implications of your work on non-target species via habitat modification, removing or interfering with certain animals within the food web, by accidental trapping, or other issues. Think through the potential outcomes in advance.
- You must comply with the requirements of the Institutional Animal Care and Use Committee (IACUC) of your research institute (Fig. 9.2). These committees prescribe the methods and standards by which you may interact with animals in a research capacity (Sikes et al., 2016). You may need to meet with the committee or a representative before commencing field research, but in all cases, must comply with their specifications and complete required documentation. In addition to state permits, you will also likely need to complete institutional applications and documentation in advance and after a field project. Be diligent and thorough in completing this

Fig. 9.2 Take care when handling, measuring and releasing wildlife. You should seek suitable training and always adhere to IACUC protocols. *Source:* Krista Keels.

paperwork as it will help the IACUC to make informed and helpful decisions regarding your work. The committee will require details including your methods statement, training, risk assessment, permits, etc. Detailed discussion of the role and responsibilities of IACUCs is provided by Sikes et al. (2016). These committees protect both you the researcher and the animal.

- Seek training before attempting any trapping or monitoring regime (Figs. 9.3 and 9.4). Training in both the practical aspects and legislative requirements will help you immensely. Your research institute may provide instruction, or you can seek out consultants or training agencies. The National Institutes of Health Office of Laboratory Animal Welfare provides online webinars and training videos that may be helpful as does the Collaborative Institutional Training Initiative Program. Remember – training in advance will save you time in the field.

Always equip live catch traps with:

- Shelter/cover
- Food
- Water

Fig. 9.3 These researchers are carefully taking measurements of various species. Training in animal handling is essential to protect the welfare of animals being studies. *Source:* Krista Keels, Katie O'Reilly, and Nikki Roach.

Fig. 9.4 These researchers are using a Fyke net to trap fish. Methodologies for fish trapping are discussed in Hubert et al., 2012. *Source:* Katie O'Reilly.

There should be enough of each of these to care for the animal for at least 24 h. Keep in mind that in adverse or harsh weather, this burden of care is greater and also, species with fast metabolisms require more frequent inspection to prevent suffering or harm.

Always check traps at least every 24 h. In many states and countries, this is the law. You must check traps in person. Devices that remotely monitor traps are not wholly reliable and failure to tend to a trapped animal may cause suffering. *You* are responsible for any and all animals you trap. If you cannot check traps in the forthcoming 24h period, you must close and secure the trap so that it cannot be triggered. A trapped animal is vulnerable to exposure, predation, and harassment by other animals or interference by humans.

Always have a backup plan if access is compromised. For example, if you are operating in difficult terrain, do you have a vehicle that can reliably get you to the trap location in all weather conditions? Or if you are operating on a farm and there is a quarantine imposed, do you have a plan for safely and legally accessing that site to close traps? Devising these contingencies is a good mental exercise before initiating a trapping regime. In the latter scenario, I would contact my local wildlife ranger and the Department of Agriculture biosecurity division and make arrangements to be accompanied under their authority into and away from the site. These are just two possible scenarios, but many others can and do occur.

All traps should be tagged with an I.D. and with a contact number. This will allow you to be contacted in the event of an accident and for the trap to be correctly identified to you so that you can respond. It is sensible to tag the location of your trap using GPS when you set it. Many commercial and amateur trappers use hunting software or apps such as ON X Maps and other proprietary programs. These can be very user-friendly, but you can also simply drop a pin on a handheld GPS (Fig. 9.5).

Position your trap correctly within the landscape. For many mammals, travel routes are ideal as some species are reluctant to enter devices positioned near their nests or dens. Look for trails

Fig. 9.5 Handheld GPS devices like these have many applications in field research, including noting locations of traps, observations, parking or camping areas and routes. *Source:* Krista Keels.

through vegetation, snagged fur and droppings. If you are experienced, you may also identify such trails by scent. This is particularly true for species that spray.

Use a correctly sized trap. Traps that are too small may exclude mature members of the target species and lead to a preferential sampling of juvenile animals.

Trapping will *never* capture all members of a population (Drickamer et al., 1999). Do not assume that your sample set is representative of the population. Individuals exhibit greater or lesser preference for traps depending on their age, past experience, physical health, rank within a pack, etc. Animals that you trap are at best representative of that portion of the population willing to be trapped via self-selection.

A failure to trap a certain species or demographic within a population does not indicate that they are absent from the area. The location in which you set the trap and your skill in setting it will influence the catch.

Non-target species can be trapped, and pets are likely to be inquisitive and lacking in the inherent suspicion exhibited by wild animals. Non-target species should be released as soon and as safely as possible.

You should *never* allow untrained individuals or third parties to interact with traps. Many people will attempt to visit traps out of curiosity, to release both target and non-target animals or to deliberately sabotage trapping endeavors. Keep in mind that trapping may be negatively perceived by some individuals, even when it is conducted for conservation or ecological research. It is wise to be discreet.

Position cage traps on a flat, level surface. Uneven surfaces will cause the trap to shift under foot (or paw) which will make the animal reluctant to enter and can lead to accidental triggering of the trap. Remember, if your trap is triggered without catching the target animal it is effectively out of commission until you can tend to it. This is a common cause of missed trapping opportunities.

Secure your traps so that they cannot be accidentally shifted (e.g., by weather) or carried away by animals. You can use a metal or wooden stake driven into the ground and attached to the trap via a chain or cable. Alternatively, you can use cable ties to attach it to a tree, fence, or post.

Many animals (e.g., Norway rats) exhibit neophobia (fear or distrust of new objects). These animals will be reluctant to enter traps until they have become acclimatized to their presence. Leaving an unset trap in an area prior to initiating your regime allows it to become weathered, and less distinct from the surrounding environment. This will increase the success of trapping once it is initiated.

Traps typically use some sort of attractant such as bait (food), scent, or lures to encourage the animal to interact with them. Just as you may have preferences for certain foods, so too will your target animal. This further confounds trapping studies, as it relates to the representativeness. A lure is anything that draws an animal toward a location. Lures can be visual (such as a flapping piece of fur), audible (such as a recorded mating call) or scent-based. Scents are usually odor-releasing substances often made from the urine, glands or fat of animals. These fool the target animal into believing that prey, potential mates, or a competitor of the same species is present or has been in the area, leading them to investigate. It is essential to consult your local or national legislation as the use of scents and lures or certain varieties thereof are prohibited in some regions and countries. In some circumstances, exemption, license, or derogation may be granted for scientific purposes.

Your local wildlife management authority may offer information as to the most effective approaches and relevant legislation.

Be aware of public perception. Some individuals and motivated groups may have preconceived notions about animal trapping that do not reflect the low risk to target species when live-trapping is conducted correctly and in accordance with the law. Perceptions of animal research have been found to vary depending on demographic (age, sex, urban vs. rural areas) and on the species of animal (Ormandy and Schuppli, 2014). Research based on live-catch trapping is integral to many conservation efforts. However, you should be aware that strong reaction to trapping can occur and should always be discreet and obey the law completely.

Euthanasia may not be an essential part of your research. Indeed, you may seek to preserve the health and well-being of the animal and ensure its safe release and in most live-catch trapping it is unnecessary. However, if you are interacting with either wild or laboratory animals, euthanasia is something that you must consider prior to beginning fieldwork. The IACUC will require you to plan for eventualities in which euthanasia is essential to prevent undue suffering of the animal or where release of a diseased animal might risk infection of other members of the community, other animals, or humans. The appropriate method of euthanasia will differ depending on the species in question; however, in all instances it should be rapid, reliable, safe, minimize pain and distress to the animal, avoid emotional distress of the researcher, and be completely legal. Before you begin any animal research consider (i) do you have the training and ability to euthanize animals if it is necessary, and (ii) have you considered possible emotional or moral concerns it might have for you personally, your team or any involved third parties? You must also plan for the safe handling and legal disposal of any resulting animal carcasses. A comprehensive guide to euthanasia protocols and related issues in research and veterinary settings is provided by the American Veterinary Medical Association (AMVA, 2013).

Use of Trail Cameras

Trail cameras (Fig. 9.6) are used both in wildlife and livestock research (Fig. 9.7). Cameras can record footage or simply take snapshots. While some cameras run continuously, it is more common to use devices that are triggered by movement. This is more power efficient and also necessitates

Fig. 9.6 Trail cameras like this one can be used in many different locations and applications. They are relatively inexpensive, sturdy and easy to use. *Source:* Sara Vero.

(a) (b)

(c) (d)

Fig. 9.7 Camera trapping is used in an enormous variety of settings and with regard to diverse species. Here are just a few. *Source:* Sian Green.

less trawling through extensive footage. These cameras use low-powered laser beams that act as a switch to turn the camera on when broken. The detection area in which these beams are effective will vary depending on the model, so always check the manual and position accordingly. Likewise, the sensitivity to movement and speed of the camera trigger will differ. Spend some time experimenting with this using a team member to trigger the camera. Cameras can be mounted on a tripod or can be strapped or screwed securely to a branch or post (Fig. 9.8). Most trail cameras are weatherproof, but keep in mind that water or dirt on the lens can render them ineffective. Just like any piece of field equipment, they should be inspected and maintained on a regular, scheduled basis. Batteries in particularly may require replacement, especially if a large amount of footage has been captured. Cameras typically save images or footage to a memory card however; some devices will transmit directly to a webpage or to your phone. If this is the case, make sure to test if it is transmitting during the installation as remote areas may have poor or non-existent signal.

Some applications of trail cameras include animal counts, investigating feeding patterns, determining predator behavior, and identifying rare animals (O'Connell et al., 2011). Some major advantages of cameras is that they have a very low or negligible effect on the animals being studied and so present a non-invasive approach and also may capture a more reliable estimate of abundance (Silveira et al., 2003). Even if you are not directly studying ecology, trail cameras can have both research and technical applications. An example of the former is in relation to understanding stream sediment dynamics. Terry et al. (2014) monitored stream turbidity at 15-min resolution at the outlet of an agricultural catchment. Trail cameras installed along the watercourse captured incidences of livestock access for drinking and by relating the timing of these events with the arrival and magnitude of turbidity, the researchers were able to disentangle sediment transport factors. A technical application is to monitor other equipment in cases of vandalism or interference.

Fig. 9.8 Trail cameras can be secured to trees, tripods or other secure locations. Always record the GPS location of your camera so it can be retrieved. *Source:* Krista Keels.

References

American Veterinary Medical Association (AMVA). (2013). *Guidelines for the Euthanasia of Animals.* Schaumburg, IL: American Veterinary Medical Association.

Drickamer, L.C., Feldhamer, G.A., Mikesic, D.G., and Holmes, C.M. (1999). Trap-response heterogeneity of house mice (*Mus musculus*) in outdoor enclosures. *Journal of Mammalogy* 80(2), 410–420.

Mulcahy, D.M. (2017). The animal welfare act and the conduct and publishing of wildlife research in the United States. *Institute for Laboratory Animal Research Journal* 58(3), 371–378.

O'Connell, A.F., Nichols, J.D., and Karanth, K.U., (eds). (2011). *Camera Traps in Animal Ecology.* Amsterdam: Springer.

Ormandy, E.H. and Schuppli, C.A. (2014). Public attitudes toward animal research: A review. *Animals* 4(3), 391–408. doi:https://doi.org/10.3390/ani4030391

Powell, R.A., and Proulx, G. (2003). Trapping and marking terrestrial mammals for research: integrating ethics, performance criteria, techniques and common sense. *Institute for Laboratory Animal Research Journal* 44(4), 259–276.

Russow, L-M. and Theran, P. (2003). Ethical issues concerning animal research outside the laboratory. *Institute for Laboratory Animal Research Journal* 44(3), 187–190.

Sikes, R.S. and the Animal Care and Use Committee of the American Society of Mammalologists. (2016). Guidelines of the American Society of Mammalogists for the use of wild mammals in research and education. *Journal of Mammalogy* 97(3), 663–688.

Silveira, L., Jácomo, A.T.A. and Diniz-Filho, J.A.F. (2003). Camera trap, line transect census and track surveys: a comparative evaluation. *Biological Conservation* 114(3), 351–355.

Terry, J.A., Benskin, C., McW, H., Estoe, E.F. and Haygarth, P.M. (2014). Temporal dynamics between cattle in-stream presence and suspended solids in a headwater catchment. *Environmental Science: Processes and Impacts* 16, 1570–1577.

The Animal Welfare Act. (1996). Public Law 89-544 Act of August 24, 1966. 89th Congress, H. R. 13881.

Printed and bound by CPI Group (UK) Ltd, Croydon, CR0 4YY

27/10/2024

14580274-0001